Your Genes and Your Destiny

A New Look at a Longer Life
when heart disease, high blood pressure,
diabetes or obesity is a family affair

by

Augusta Greenblatt
and
I. J. Greenblatt, Ph.D.

with a foreword by Arno G. Motulsky, M.D.

The Bobbs-Merrill Company, Inc.
Indianapolis/New York

Other Books by Augusta Greenblatt

Heredity and You:
How You Can Protect Your Family's Future
Teen-age Medicine:
Questions Young People Ask About Their Health
Why Do I Feel This Way?
(paperback revised version of *Teen-age Medicine*)

Library of Congress Cataloging in Publication Data
Greenblatt, Augusta.
    Your genes and your destiny.
    1. Medical genetics. 2. Environmentally induced diseases.
I. Greenblatt, Irving Jules, 1912–, joint author. II. Title.
RB155.G76   616'.042   77-15439
ISBN 0-672-52302-7

For Ricky and Susan, and Larry

Contents

foreword

The last few years have seen much interest in genetics. The media are full of stories about bizarre scenarios featuring recombinant DNA and cloning. Unfortunately, the public press and TV give less attention to recently introduced and current applications of medical genetics, such as genetic counseling, genetic screening, and intrauterine diagnosis, which are less sensational. These medically oriented activities involve families and patients who are affected or are at risk for the many different genetic diseases. These topics were covered in Augusta Greenblatt's earlier book *Heredity and You*.

Her new book *Your Genes and Your Destiny*, with coauthor I. J. Greenblatt, Ph.D., deals with more complex but far-reaching issues of medical genetics. Many of the common diseases of middle life, such as diabetes, high blood pressure, and coronary heart disease, show familial aggregation. The high frequency of these disorders makes them important public health problems. Unlike single-gene disorders (such as hemophilia, sickle cell anemia, or Tay Sachs disease), in chromosomal aberrations (such as Down syndrome — mongolism), the underlying genetic basis has only been identified in a few instances. Nevertheless, various studies designed to tease out the relative roles of heredity and environment have shown that the observed familial aggregation is based on heredity and cannot be accounted for by a common familial environment.

The data are usually interpreted to indicate that several genes rather than a single gene or chromosomal aberration are involved. Intensive work into the genetics of many of these disorders is just beginning and is usually difficult. In most instances, it has become clear, however, that the yet unknown genes alone are not sufficient to bring about the disease. Interaction with environmental factors — sometimes known, often unknown — are required to cause disorders such as hypertension or coronary heart disease.

The authors stress the many scientific leads that are being studied by biomedical researchers all over the world. Their book attempts to show that the genetic contribution alone is *not* human destiny. Recognition of the specific gene-environmental interrelation in a given disease may make it possible to change an unfavorable outcome through a healthier lifestyle, diets, or certain medicines, particularly if practiced early in life. Medical geneticists and genetic epidemiologists in their work are interested to find tests and methods by which susceptible or predisposed persons can be identified for such preventive measures. The final outcome is hoped to be a preventive medicine adapted to a person's strengths or weaknesses.

My own work for many years has emphasized the relationship of genes to the environment in the causation of human disease. For example, drug reactions caused by genetically determined enzyme defects are models for the common conditions discussed by the Greenblatts. Some person with a specific enzyme defect may develop a harmful reaction when given a drug that requires the abnormal enzyme for its breakdown. The enzyme defect alone is harmless, as is the usual dose of the drug in a person without the enzyme defect. Disease (the drug reaction) only occurs when nature (the enzyme defect) and nurture (the drug) happen to interact at the same time in a person. It is likely that susceptibility to injury from environmental chemicals may also vary in different persons for genetic reasons. The field of pharmacogenetics and ecogenetics has developed from such concepts.

Work on the genetic basis of coronary heart disease initially has led us to identify genes affecting cholesterol and triglyceride metabolism which predispose their carriers to premature heart attacks. Much more research on the fundamental action of these genes is required. The work of Goldstein and Brown has clarified the basic defect of the gene for familial hypercholesterolemia, which occurs in about 1/500 of the population. The Greenblatts pay particular attention to families with this disease in their book.

Dr. and Mrs. Greenblatt's account covers many different research leads, with stress on the latest and newest results. Their book succeeds beautifully in portraying the excitement and hope resulting from work in this area. Application of the insights of modern genetics to our emerging understanding of the fundamental mechanisms of disease, development and aging promises to yield high dividends in better health.

The success of this book in educating the public toward better understanding of the thrust of modern research in some areas of medical genetics and related fields will be a sane antidote to occasionally sensationalist press accounts which often leave a wrong impression of the interests and motivations of genetic researchers.

Arno G. Motulsky, M.D.
Director, Center for Inherited Diseases
Professor of Medicine and Genetics
University of Washington
Seattle, Washington

Introduction

This is the story of a medical "breakthrough" that meets none of the usual criteria. It makes no spectacular headlines, involves no magic medication, and promises no overnight cure. It does, however, show the way to a longer life for the millions threatened by the not-so-rare hereditary diseases.

If your father had a coronary heart attack at 48, or your mother at 60 has just learned she is a diabetic, or your brother suffered a stroke at 53, you, too, may be at risk.[1] Your genes, however, *need not be the total sum of your destiny.* Although the complex genetics involved in these disorders are just beginning to be unraveled, there is no longer any mystery about identifying who is at risk, what can be done about it, and, in many instances, how to prevent it from developing.

The good news is that coronary heart disease, the most serious epidemic in the twentieth century, is on the wane. Atherosclerosis, the underlying cause, is both preventable and reversible. Strokes and other consequences of untreated high blood pressure have declined dramatically. The enigma of diabetes comes closer to solution with the new view that it is two distinct disorders. And there is a growing consensus that aging and disease are not always destined to go hand in hand. A crucial ingredient in this

---

[1] One in every five coronary victims under 60 has a hereditary predisposition. Among the 25 million with high blood pressure, heredity plays a role in 20–30 percent of blacks, and 15–20 percent of non-blacks. The vast majority of the 10 million diabetics have inherited their vulnerability.

success story is you and the lifestyle you adopt to protect yourself and your family, for while you cannot change your genes, you *can* control the environment in which the genes flourish.

With the completion in 1974 of *Heredity and You: How You Can Protect Your Family's Future*, we felt, to some degree, as if we were waiting for the other shoe to drop. While the book dealt in detail with genetic counseling, Tay Sachs, sickle cell, Down syndrome (mongolism), Rh, hemophilia, cystic fibrosis, thalassemia, mental illness, IQ, Huntington's disease, pharmacogenetics (inherited response to drugs), Wilson's disease, PKU, and other inborn errors, it was clear that just a proposed chapter on the not-so-rare diseases could not adequately tell their story. The fact is that *Your Genes and Your Destiny*, which focuses only on CHD, high blood pressure, diabetes, obesity and aging, could not have been written in 1974. The few short years since 1974 have witnessed an exciting proliferation of new knowledge which we have painstakingly followed and chronicled.

Physicians have a new awareness of inherited vulnerability, with the spotlight shifting to the young child and adolescent at risk. Pediatricians are routinely recording blood pressures in patients over 3 (a family trend is discerned early in life), and family doctors are looking at sons and daughters of CHD victims in their forties and fifties.

The study of genetic diseases is gaining increasing prominence in health education. Seattle high school students will soon be learning about genetic counseling, Tay Sachs, sickle cell, Down syndrome, and cystic fibrosis. What started in California in 1974 as a special project on sickle cell disease has been expanded by the State Education Department into a comprehensive course of study on human heredity and birth defects, and by 1978 was ready for use in public schools throughout the state from kindergarten through adult education. High schools and colleges

across the country are using the National Foundation–
March of Dimes filmstrip "From Generation to Generation: Genetic Counseling," which not only alerts students
to the possibility of transmitting a specific genetic disease,
but also helps in finding local sources for counseling.

Prevention is traditionally one of the least glamorous
disciplines in medicine, but the newly demonstrated
promise of a longer life — even in the absence of an
inherited vulnerability — has a glamour all its own.
Supermarket carts are filled with foods less likely to promote atherosclerosis. Leisure activities are more likely to
include jogging and tennis, and men are likely to be smoking less. And for the first time, an innovative program in
prevention involves junior high school students in a
number of school districts in the New York metropolitan
area who are taking responsibility for their own future
health by committing themselves to a lifestyle that will
help prevent heart disease, high blood pressure, obesity
and cancer.

We would like to thank the dozens of scientists and
clinicians in the vanguard of the field who shared their
thoughts and hopes for the future with us. To be sure,
with the wide panorama we covered, there are inevitably
points of view expressed that do not find unanimous
agreement among all the researchers. All, however, were
unanimous in their candor in admitting how much was
still to be learned, and all shared our enthusiasm for the
timeliness of this book. We are also grateful to the following for their cooperation in providing material, as well as
for the time they took to review some of the chapters. Any
mistakes are ours alone.

Leatrice Ducat, Founder, Juvenile Diabetes Founda-
tion, member National Diabetes Advisory Board; Stanley
M. Garn, Ph.D., Professor of Human Nutrition and of
Anthropology, The University of Michigan, Ann Arbor,
Michigan; Richard J. Golinko, M.D., Director of Pediatrics

and Pediatric Cardiology, Brookdale Hospital Medical
Center, and Professor of Clinical Pediatrics and Director of
Pediatric Cardiology, State University of New York,
Downstate Medical Center; Irwin Kanarek, M.D., Chief of
Ophthalmology, Brookdale Hospital Medical Center,
Brooklyn, New York; Richard S. Levine, M.D., Co-Chief,
Cardiology and Physician in Charge, Cardiac Catheteriza-
tion Lab, Brookdale Hospital Medical Center, Brooklyn,
New York; Robert I. Levy, M.D., Director, National
Heart, Lung and Blood Institute, National Institutes of
Health; Arno G. Motulsky, M.D., Director, Center for
Inherited Diseases, Professor of Medicine and Genetics,
University of Washington, Seattle, Washington; Lot B.
Page, M.D., Chief of Medicine, Newton-Wellesley Hospi-
tal, and Professor of Medicine, Tufts University School
of Medicine, Boston, Massachusetts; American Heart
Association; American Diabetes Association; Juvenile Dia-
betes Foundation; National Institute of Arthritis, Metabo-
lism and Digestive Diseases; National Eye Institute; Na-
tional Heart, Lung and Blood Institute; National High
Blood Pressure Education Program; National Institute on
Aging; Gerontology Research Center, Baltimore, Maryland.

Closer to home, a special thanks to the first reader of the
work — Corinne Boni — who met every typing deadline
with skill and accuracy; our appreciation, too, for the
services and facilities available to us seven days a week at
the Hewlett-Woodmere Public Library.

Finally, we owe much to our editor, Barbara Norville,
for her unflagging interest, encouragement and insight.

Augusta Greenblatt
I. J. Greenblatt, Ph.D.
Woodmere, New York
October, 1978

1

Understanding Heredity

It was bound to happen sooner or later. An aging millionaire, unmarried and childless, decides that he wants a son — an exact replica of himself, bypassing the genetic contributions of a mother to the baby. He has probably read that it has been done in frogs, and he may even have seen it satirized on the screen in a hilarious attempt to duplicate a dead dictator from his nose.[1] What worries scientists today about the still unproved story[2] of a human cloning is not whether it is fact or fiction, but that such a sensational claim, with the moral and ethical dilemmas it poses, can create a backlash that will seriously hamper basic research and human genetics.

Scientists who speculated a decade ago about cloning a copy of your favorite person scoff at the idea today. Many are appalled at the purported use of genetic research to gratify the bizarre desire of an egotistical millionaire, and although the consensus is that it is theoretically possible,

---

[1] Woody Allen's *Sleeper*.

[2] Related as fact by David Rorvik in *In His Image — Cloning of a Man* (Lippincott, 1978). Rorvik, however, acknowledges that he has no direct proof that the father and son are genetically identical.

there are still too many unsolved technological problems for it to be a probability today.

Will it be accomplished in the future? A more realistic question is "Who needs it?" The fact is, one of the most precious parts of our heritage is our genetic diversity and variability. Even bacteria, when they exchange genetic information before dividing, experience sexual reproduction in the most basic sense.

All new life begins with a single cell, whether it is the one-celled bacterium dividing to produce two daughter cells virtually indistinguishable from the parent, or the fertilized human egg that will develop into an individual with billions of cells, each with highly specialized functions. The finished product — you — will bear a resemblance to your parents, your grandparents, assorted aunts, uncles, sisters and brothers, but will still be uniquely you. Indeed, so numerous are the variations possible from the genetic information transmitted by your mother and father, that unless you have an identical twin, you are probably genetically different from any of the 70 billion individuals who have peopled the earth.

The real news today is in the impressive strides made in the last ten years as medical genetics moved out of the laboratory and into clinical practice. While there are still no cures for the wide array of genetic disorders — some striking thousands, others threatening millions — an increasing number can be prevented, treated and controlled.

There is news in a new technique of experimental genetics called gene splicing (recombinant DNA), which promises to "put us at the threshold of new forms of medicine, industry, and agriculture."[3-4]

---

[3] Testimony before the Senate Subcommittee on Commerce, November 1, 1977, by Dr. Philip Handler, President, National Academy of Sciences, and Dr. Paul Berg, Stanford University.

[4] In all, eighty-six universities and nine industrial companies are now

☐ In September 1978 a team of West Coast research scientists working at the City of Hope National Medical Center, Duarte, California, and Genentech, Inc., announced that they had succeeded in synthesizing the genes for human insulin, spliced the instructions into the genetic machinery of a common laboratory bacterium, and "commanded" the bacterium to produce human insulin. For millions of diabetics around the world dependent on insulin for their survival, this landmark accomplishment promises a continuing supply of the human hormone in place of the diminishing supply of insulin now extracted from cattle and pigs.

☐ Less than a year earlier (December 1977), the same team, working with scientists at the University of California, San Francisco, had scored a historic first — the laboratory production of a recently discovered brain hormone, somatostatin[5] — using the same technique of gene splicing that was later to produce human insulin.

☐ A giant industrial company has won the right to apply for a patent on a bacterium that has been given a gene with information to devour oil — with the potential to clean up oil spills in coastways and oceans.

☐ The hope for the future is that crops that now need fertilizer for optimum growth (such as wheat and corn) can be given genetic information to make their own fertilizer by trapping nitrogen in the atmosphere, as soy beans, peas and other nitrogen-fixing plants already do naturally.

For many of the victims of the more than 2,000 known genetic disorders and their families, however, the future is here today. For the first time in history, there are answers

---

involved in recombinant DNA research. Federally funded projects are governed by NIH guidelines. Legislation to regulate *all* research is expected in 1978.

[5] See Chapter 4.

and options for the anguished parents who ask, after the birth of a child with a severe genetic disease, "Why did it happen to us?" and "Can it happen again?" There are reassurances for many who may not have yet suffered the tragedy of an affected child but are at risk because a close relative has; or because they are members of an ethnic group in which specific genetic disorders are known to cluster — Tay Sachs disease in Jews of eastern European ancestry; thalassemia in Greeks, Italians and Asians; sickle cell disease in blacks. There is hope for the child born with a rare metabolic disorder such as PKU or galactosemia where early recognition and an appropriate diet instituted in time can save a baby otherwise doomed to mental retardation and other problems. And there is new reassurance for the mothers-to-be over 35 who account for a disproportionate number of babies with mongolism (Down syndrome).

Now, for the first time, the focus is also on the not-so-rare hereditary diseases, not as drastic or dramatic in their manifestation, but contributing to premature death or disability of millions. For every black who will die of sickle cell disease, a hundred will die of hypertension; and perhaps as many as 30 percent of that hundred will have inherited their vulnerability. It has long been known that diabetes runs in families, and today there are new insights into who is at risk and what can be done about it. Coronary heart disease as a family affair is not new in medical annals.

One such family was that of Matthew Arnold, nineteenth-century British critic, poet and novelist, who was the third generation in his family to die suddenly of a heart attack. When he went to meet his daughter and granddaughter on their way home from America, "In his joy . . . at seeing them so soon," a friend wrote, "he leapt over a low fence, and alas! dropped down dead." His father, Thomas, a famous educator and religious philosopher, also died suddenly from a heart attack one day before he was to celebrate his 47th birthday, but not before he had informed a particularly perceptive and probing doctor that *his* father before him, a customs inspector, had also died suddenly while still a young man, leaving 6-year-old Thomas an orphan.

It is only in very recent years that specific genetic defects have been associated with these complex disorders often unsuspected and undetected until adulthood. Usually, more than one gene is involved — and always with a tangled interaction of genes, environment and lifestyle. "Human genetics is a very young field," says Dr. Leon Rosenberg, Chairman of the Department of Human Genetics, Yale University School of Medicine, "born in this century, infantile until twenty years ago, and now undergoing a most impressive adolescent growth spurt."[6] It still has a long way to go to reach maturity, but there is no doubt that slowing down the growth of new knowledge and application of existing knowledge, because of a hostile and suspicious climate, can only cause needless pain and suffering.

Unraveling the mystery of heredity is one of the most suspenseful and far-reaching stories in science and medicine. When Hippocrates observed in the fourth century b.c. that "bald people are descended from bald people, people with blue eyes from people with blue eyes, squinting persons from squinting persons," he was not the first to recognize that living things can pass on their characteristics to their offspring. As far back as 10,000 years ago, Stone Age man learned that certain strains of barley could be depended upon to yield vigorous and nutritious grain, crop after crop.

---

[6] Foreword to *Heredity and You: How You Can Protect Your Family's Future*, by Augusta Greenblatt (New York: Coward McCann and Geoghegan, 1974).

In time, man's interest in heredity turned to his own biological origin. Only three hundred years ago, one popular theory placed a fully formed little man or woman (homunculus) in the egg which, when triggered by the sperm, would start growing. Another view held that the little creature was in the sperm and was injected into the egg at conception. While we chuckle at the naïveté, the correct explanation is even more incredible. All of the information that will make you uniquely you is in a single cell to which each of your parents has contributed equally. It contains enough information to fill 46 volumes (23 from each) of 15,000 pages each.

DNA,[7] the substance in which the information is coded, is deceptively simple. The language of heredity uses a four-letter alphabet (AGCT[8]), three-letter words and a universal code for all forms of life — one-celled bacteria, fruit flies, sharks, mice and man, and has probably been in continuous use for more than three billion years. DNA is an extremely long, thin molecule, occupying so little space that the genetic information of the world's population today would easily fit into the tip of a teaspoon. It transmits information with astonishing accuracy from generation to generation, and from day to day within your own body, as old cells are replaced by new and as it directs the machinery for carrying on thousands of complex chemical processes. If a good typist were to copy the massive set of instructions in our genetic endowment, one could expect one mistake in every 20 pages. In nature, the mistakes (mutations) take place about one in every 10 million pages. In rare instances, the mutation is a good change, conferring an advantage. This is how evolution of new forms and new characteristics comes about. Sometimes the mutation is trivial — each of us has six to

[7] Deoxyribonucleic acid.
[8] Adenine, guanine, cytosine, thymine.

ten defective genes that impose no hardship. But some mutations are sufficiently serious to threaten health and survival and account for the wide range of genetic diseases.

DNA was first described more than a century ago by a young Swiss biochemist — Friedrich Miescher. By that time, scientists already knew that the cell is the basic unit of life, and that all living cells come from other living cells. They also knew that a dense demarcated region in the cell, the nucleus, was the seat of heredity and of all the cell's other functions. Remove the nucleus and the cell no longer lives, grows or divides. Miescher isolated two substances from the nuclei of white blood cells and salmon sperm — a protein fraction and a non-protein rich in phosphorous, which turned out to be DNA. In a choice of which held the secret of life in the nucleus, the protein was a natural front-runner. Proteins are very complex compounds made up of hundreds of amino acids and account for more than half of the structures in our body — skin, muscle, blood, etc. When DNA was later analyzed, it turned out to be relatively simple — a molecule of sugar, a molecule of phosphate, and four molecules identified as adenine, guanine, cytosine and thymine.[9] There was no logical evidence at the time to attribute to it the blueprint of life.

Contrary to the popular belief in the primacy of protein, Friedrich Miescher clung to his faith in DNA. He knew from his observations of salmon during their long migration from the sea to their spawning grounds, when they abstained from food completely, that they nevertheless continued to manufacture large amounts of nuclear material from precursors in their own bodies. But seventy-five years were to pass before he was proved right. In fact,

---

[9] RNA, a substance similar to DNA but differing in minor respects, was also identified at a later date. Some viruses have their hereditary data in RNA.

when we were in graduate school in the late 1930s, we were still learning that the nucleic acids were ubiquitous both in plants and animals, but always downgraded in importance because of their simple structure.

Miescher won long-delayed recognition in 1944, when Dr. Oswald T. Avery and his colleagues, Drs. Colin MacLeod and Maclyn McCarty, of the Rockefeller Institute, proved that when harmless Type 2 pneumococcus bacteria were transformed into virulent, lethal Type 3, the DNA, not the protein, carried the information for the transformation. It was a landmark discovery, and although it did not win a Nobel prize for the team, it sparked a new era in genetic research, attracting a new type of scientist — the molecular biologist — using the tools of physics, mathematics, biochemistry and biology, as well as the techniques of classical genetics.[10]

By 1953, there were enough facts known about DNA waiting to be put together to explain it. X-ray pictures by a brilliant young woman, Dr. Rosalind Franklin of Kings College, London, suggested it was a double helix, like a spiral staircase. Columbia University's Dr. Erwin Chargaff discovered how four simple compounds (the bases AGCT) could be lined up to provide the myriad different sequences in every living thing. At Cambridge University, Dr. Francis Crick and Dr. James Watson, a young American who had sought an appointment there with the expressed goal of being the first to explain DNA, were working with models that they hoped would put it all together to provide a complete picture. They were waiting

[10] Winning Nobel prizes for George Beadle, Edward Tatum, Joshua Lederberg, Arthur Kornberg, Severo Ochoa, Francis Crick, James Watson, Maurice Wilkins, Francois Jacob, Andre Lwoff, Jacques Monod, Marshall Nirenberg, H. Gobind Khorana, Robert W. Holley, Max Delbruck, Alfred Hershey, Salvador E. Luria, Renato Dulbecco, David Baltimore, Howard M. Temin, Hamilton Smith, Daniel Nathans, and Werner Arber.

for a new model to arrive when they had that an-swer[11] — hypothetical to be sure — but soon tested out by excited experimental scientists.

Their hypothesis stood up. DNA is a double helix — two long fibers coiled around each other like a spiral staircase, with a backbone of alternating molecules of sugar and phosphate. The four bases, AGCT, are the rungs of the ladder, and in order for the rungs to fill the spaces between the sides of the ladder properly, A must be paired with T, and G with C. The attachment is very weak, so when DNA starts to make a copy of itself, the coils unwind into two separate single strands and a new strand forms alongside the old. An A picks up a T, and a G picks up a C from the nucleotides[12] in the cell. There are now two new double helixes, identical to each other and to the original, and each is made up of one old strand and one newly constructed — and perfectly complemen-tary — strand. The speed and ease with which this hap-pens is mind-boggling. In minutes, a chain of 200,000 nucleotides unwinds without tangling, picks up a new partner, and winds itself into a double helix again.[13] Rep-lication has been accomplished.

In a sense, the advocates of proteins as the basic mole-cules of life were right. A principal function of DNA is to direct the production of proteins, including the hemo-globin in your red blood cells that carries oxygen to your cells, and the enzymes that catalyze metabolism and other chemical reactions, the hormones that control growth, reproduction, etc. You do not inherit a gene for skin color.

[11] *Nature*, Vol. 171, 1953, pp. 737–38.
[12] Consisting of one of the bases (A, G, C or T), a sugar and a phosphate.
[13] A few years later, Dr. Arthur Kornberg, then at Washington Uni-versity (now at Stanford University), would replicate the feat in the laboratory and win a Nobel prize for doing so.

More accurately, you inherit at least nine genes that, in turn, direct enzymes to dictate the formation of pigments that will show up as the color of your skin. We have about 100,000 genes made up of 10 billion nucleotides, with enough information to code for 10 million chains that will be assembled to make up our proteins.[14] When a specific protein is needed, only the segment of the DNA with the appropriate gene becomes involved. It acts as a template for a closely related molecule called messenger RNA, which leaves the nucleus with its message and proceeds to a structure in the cytoplasm — the ribosome — where the actual assembly of the protein takes place. Still another RNA — transfer RNA — selects the appropriate amino acids in the cell and brings each one back to the ribosome to fill the order.

How can DNA, with its four-letter alphabet, dictate the selection of the correct amino acids among the twenty available,[15] and follow through to have them assembled in the proper sequence into a protein molecule with its long chains made up of hundreds of amino acids? The code was broken by a young NIH scientist, who revealed it at the 1961 Fifth International Biochemical Congress in Moscow. Dr. Marshall Nirenberg found that with only three nucleotides of DNA and messenger RNA, every amino acid for every protein can be specified. He broke the code for one amino acid, and in the next few years, the codes for all the other amino acids were also worked out. A mutation, or change, in heredity could now be explained by a change in a single molecule.

Making a protein is an extraordinarily well-orchestrated event, where a single wrong note will continue to be heard

[14] At any one time only a very small number of the genes are active.
[15] You can make about half of the 20 amino acids you need. The others come from the plant and animal protein in your diet.

from. If one portion of DNA (a gene) is not duplicated properly, all the daughter cells will bear the same mistake, which will then be repeated in the RNA. The code will dictate the wrong message, and the wrong amino acid will be placed into a protein. *A single incorrect amino acid* among hundreds of correct ones in a protein chain can have far-reaching, even fatal, consequences. So it is with the Tay Sachs baby, seemingly healthy and beautiful at birth, but who begins after six or seven months to show signs of the deterioration that will culminate in blindness, paralysis and a vegetable-like existence, with death by the age of 3. The defect has been tracked down to a single gene that dictates the production of a specific enzyme — hex-A[16] — whose function it is to break down a specific fat. Without the enzyme, the fat, which is produced at a normal rate, accumulates in the brain in abnormal amounts. The damage is irreversible and inexorable.

Sickle cell disease is a hereditary disorder that claims for most of its victims inhabitants of West and Equatorial Africa and American blacks of African ancestry, as well as non-blacks in India, southern Arabia, and sections of Mediterranean Europe. It means a shortened life span punctuated by bouts of anemia, infection, pain and weakness. It is also the first disease linked to a defect in a single protein molecule — hemoglobin — which, in turn, is linked to a single defective gene.[17] The original mistake (mutation) in the DNA must have happened thousands of years ago, but it is only with the cracking of the genetic code in the 1960s that scientists learned the basis for the mistake. Among the hundreds of amino acids that make up the protein fraction of hemoglobin, one is incorrect.

[16] Hexosaminidase A.
[17] "Sickle Cell Anemia: A Molecular Disease," Dr. Linus Pauling, *et al*. *Science*, November 25, 1949.

The code for glutamic acid (correct) is strikingly similar to the code for valine (incorrect).[18]

Genes are no longer a complete mystery. It is a piece of hereditary information, a chemical entity whose structure and modus operandi are known. Genes have been photographed at rest and at work. There is now evidence of their versatility — a single gene can dictate instructions for more than one protein. And late in 1976, Nobel Laureate Dr. Har Gobinda Khorana and his colleagues at MIT succeeded, after nine years' work, in putting together an artificial gene — complete with stop and go signals — that works.

While Miescher was studying the chemistry of the nucleus, other scientists of that era were looking at its structure through their microscopes. With the development of organic dyes in the 1880s, new and important clues began to emerge when the cells were stained and examined under the microscope. When the cell is ready to divide, the random jumble of thread-like bodies in the nucleus shapes up into discrete bodies (chromosomes), the bodies line up, and each makes a copy of itself. When division (mitosis) is complete, each of the daughter cells has the same type and same number of chromosomes as the mother cell. Different species have different numbers of chromosomes — mice, 20; monkeys, 42; man, 46 (more precisely, 23 pairs). There is no relationship between the number of chromosomes and status in the evolutionary ladder (some crabs have 500). There is an important exception to 46 chromosomes in each cell. In preparation for fertilization, the egg and the sperm undergo one cell division without replication. They thus end up with 23 chro-

---

[18] By the mid-70s, more than 140 hemoglobin abnormalities had been described. Some are relatively harmless; others, when found with the sickle gene, may increase or decrease the severity of the disease.

mosomes each and, in the process, drop half of the genetic information for that particular mating (meiosis). Additional shuffling of chromosomes takes place so that each of your four grandparents is represented in the final 46 of the fertilized egg. The variety possible with such shuffling allows for about eight million different combinations for each offspring.

Scientists' early suspicions that chromosomes were involved in heredity were confirmed in the early years of the twentieth century, in the proliferation of research generated by the rediscovery of the work of an obscure Augustinian monk. What Gregor Mendel called "a character," we call a gene — a particular point on a chromosome where a piece of hereditary information controlling a particular trait is located.

You may have heard an expression commonly used to describe an uncommonly useful object or concept: "If it did not exist, someone would have to invent it." The fact is, that is exactly what Mendel did with the gene more than a hundred years ago, in the first scientific study of heredity. He needed the "characters" to explain the results of eight years of painstaking work with 20,000 sweet pea plants in his monastery garden in Austria. He had prepared well for his research — he had spent four semesters at the University of Vienna, where he had studied physics, chemistry, botany and mathematics with some of the leading researchers of the day. His aim was to find out what happens to hybrid plants — offspring of parents with different characteristics — when they breed. What shows up in succeeding generations? Is there a pattern and can it be predicted? These are the questions we ask about ourselves.

Mendel had already started his experiments on heredity in sweet peas when an event almost a thousand miles away rocked scientific, intellectual and religious circles. In 1859,

Charles Darwin and Alfred Wallace independently announced that the varied forms of life on earth had not appeared readymade, but that various species had evolved from common ancestors by imperceptible changes over millions of generations. Darwin needed an explanation of heredity to explain not only how the rare changes came about, but also how the more usual faithful replication was ensured. The "homunculus" theory had long since disappeared. Along with many of his contemporaries, Darwin believed that each part of the body contributed its qualities to the germ cell (egg and sperm) and the offspring represented the blending of the characteristics of each parent at the time of conception.

The theory had some attractive features and even allowed acquired characteristics to be transmitted. (This was before zoologist August Weismann cut off the tails of twenty-two generations of mice and learned that each generation was born with the same size tail his ancestor had had at birth.) And it did not explain why some children resemble their grandparents, aunts and uncles more than their parents. Mendel found some important answers to Darwin's dilemma. He knew they were important, for he had read Darwin. In the introductory remarks of his 1865 paper, he explained that he had undertaken such an ambitious set of experiments in order to "reach a solution of a question the importance of which cannot be overestimated in connection with the history of the evolution . . ." Darwin died sixteen years after Mendel's work was completed without ever having seen it or ever having heard of Mendel. Indeed, Mendel's discoveries were to remain unnoticed for almost thirty-five years before they were dusted off and read with respect for the first time.

For his experiments, Mendel assembled purebred specimens with a number of visible differences — such as violet-red and white flowers, short and long stems, rough and smooth seeds. When violet-red flowers were bred with white, all the offspring in the first generation were violet-red. Where had the white gone? Mendel found out when he bred the new violet-red with one another. They produced both violet-red and white. A characteristic of the grandparents was there all the time, but it seemed to skip a generation before it showed up again. He also learned that long-stemmed violet-red when bred with short-stemmed white could produce in succeeding generations not only offspring of each color, but also long or short stems with each color. The two traits had separated from each other. His conclusion: heredity is not a blending — the characters are transmitted in discrete, independent units from generation to generation.

Much of what Mendel learned from his plants applies to us — and has been crucial in solving long-standing puzzles both in health and in disease. Each of your parents contributes equally to your inheritance, with a gene for every trait. If both genes are identical, you are homozygous for that trait. If the genes dictate a different message, you are heterozygous. Some traits will show up even in the presence of only one gene — they are dominants. Other traits need *both* genes to show up — they are recessives. This explains why a trait can be present for generations and not show up. It also explains how you can inherit blue eyes from two brown-eyed parents. When that happens, each of your brown-eyed parents is a heterozygote, and each, purely by chance, has transmitted the blue-eyed gene to you.[19]

---

[19] Actually, inheritance of eye color is not as straightforward as the above illustration suggests. At least five genes are involved, including a masking gene which sometimes results in a brown gene not being expressed in one generation but being transmitted nevertheless. This explains the occasional phenomenon of two "blue-eyed" parents with a brown-eyed child.

It also explains how two healthy parents can transmit a devastating disease to a child. The genes for Tay Sachs, sickle cell, cystic fibrosis, thalassemia, and hundreds of other disorders must be present in a double dose, and that happens only when each parent is a carrier (has one defective gene and one normal gene).[20] In each pregnancy, chance alone determines which gene will be transmitted. The probability that each parent will transmit the good gene — and thus the child will be free of the disorder — is one in four. The probability of transmitting a good gene from one parent and a defective one from the other — and thus the child will be, like the parent, a healthy carrier — is two in four. The probability that each parent will transmit the defective gene — and thus the child will be a victim — is one in four.

But chance has no memory. These odds are the same in each pregnancy, and statistics are valid only when applied to thousands. In any particular family at risk because both parents are carriers, all the children may be normal, or all may be affected. When a disease is caused by a dominant (only one dose of the gene required), each offspring has a 50-50 chance of inheriting it from his affected parent. The other parent's normal gene is no protection. The child who inherits the affected parent's normal gene is, of course, safe from the disease. The chain of inheritance is broken. He is not a carrier and so his children are also safe.

A few years after Mendel's publication, a 22-year-old Long Island physician who had never heard of Mendel's theories accurately described for the first time the pattern of inheritance of a dominant gene in a rare destructive

---

[20] One in every 30 American Jews of eastern European ancestry is a Tay Sachs carrier; 2 million American blacks are sickle cell carriers; 5 percent of American Greeks and 2½ percent of American Italians are thalassemia carriers; an estimated 10 million Americans, predominantly white, are carriers of cystic fibrosis.

neurological disorder that bears his name. Huntington's disease has become familiar to both the profession and the public since it claimed the life, in 1967, of one of America's greatest folk heroes — Woody Guthrie.

The most recently discovered dominant is a gene that controls one aspect of cholesterol metabolism, placing the individual at high risk for premature coronary heart disease. With an incidence of 1 in 500, it is probably the most common single-gene defect in the American population.[21]

The total information in your genes is your genotype; the way the various characteristics are expressed by your genes is your phenotype. Although your genotype cannot yet be changed, your phenotype often can. Indeed, so important are environment and experience that even identical twins will often differ in height, weight, interests and susceptibility or resistance to disease. (This was not a problem to Mendel, who kept a tight control over the conditions under which his peas grew.) When diabetes, a disorder with a large genetic component, strikes a twin under 45, the risk for the other twin is no more than 60 percent. And identical twins reared apart may have an IQ spread of as much as twenty-four points. The recognition that your genes are not your destiny is the rationale for today's treatment of genetic disorders — ranging from the rare child with galactosemia who will thrive when foods containing milk sugar are removed from the diet, to the hypertensive who is saved from a stroke by losing weight, throwing away the salt shaker, and, if necessary, taking appropriate medication.

Attempts are now underway to change the internal environment directly in a number of genetic disorders by replacing missing enzymes. In the opinion of NIH re-

[21] See Chapter 2.

searcher Dr. Roscoe O. Brady, trials so far are sufficiently encouraging to warrant further investigation.[22] In one attempt to replace the missing enzyme in Tay Sachs, the level of hex-A in the blood was raised to normal, but none of it reached the brain. It was unable to breach the blood-brain barrier. For the forseeable future, there is little hope that Tay Sachs will be treatable.

There is hope, however, for the family faced with the problem today. The worst shock is when it happens for the first time in a family's memory, leaving the stunned parents fearful for the future. Why did it happen, and will the next pregnancy (if there is a next) be a blind lottery? It is no longer enough to know the mathematics of the risk. The parents of the Tay Sachs child can now be told, "Yes, we can explain to you how it happened and what went wrong in the chemistry. But more important, when you decide to become pregnant again, it need *not* be a blind lottery. We can monitor the pregnancy, reassure you if the baby is free of the disorder (the odds for that are three in four) or, if it *is* affected, let you make the choice of terminating the pregnancy, with the hope that the next will be okay."

Tay Sachs disease is one of 125 known inborn errors of body chemistry. More than 50 can now be diagnosed in utero by amniocentesis, a process in which a sample of the fluid surrounding the fetus in the womb is withdrawn from the mother's abdomen and both the fetal cells and the fluid itself are studied. The procedure is safe and accurate, and can detect not only biochemical errors, but also abnormalities in the chromosomes. The life-giving potential of the advance in genetic counseling is underscored by its

---

[22]Most encouraging was the response in Fabry's disease, where death usually occurs from kidney damage, and in the adult form of Gaucher's disease, marked by damage to liver and bone.

increasing use by parents who have not yet suffered the tragedy of an affected child but know they are at risk.

Marcia and Len were tested for Tay Sachs shortly after they were married. Both are healthy, with *no* history of a Tay Sachs child in either family, but a simple, inexpensive blood test revealed that each is a carrier. Only a decade ago, they would have had no way of knowing that there was a risk of a baby born with a rare lethal disease of which they had never heard.

To one woman who started to raise a family at 38, the opportunity to monitor her pregnancy was the determining factor in her decision to become pregnant. She was at double risk. Twenty percent of all babies with Down syndrome (mongolism), in which one chromosome too many (47 instead of 46) accounts for the mental retardation and other features characteristic of the disorder, are born to mothers over 35, who now account for no more than 10 percent of all live births.[23] Furthermore, *her* mother had had three children with Down's. In the light of her family history and age, the risk, when left to pure chance, was too great to contemplate. A test of her chromosomes revealed that she carried the same defect as her mother (a balanced translocation on one chromosome) and even if she were ten years younger, there would still be a risk.

She became pregnant, only to learn from amniocentesis that the fetus was affected. Her decision was to terminate,

[23] In the past more than 50 percent of Down syndrome babies were born to mothers over 35. In recent years, however, with older women bearing fewer children, the vast majority of affected babies are the offspring of younger mothers. Furthermore, new advances in chromosome staining reveals that one in four Down babies receives the extra chromosome from the father. Genetic counselors can now help parents better confront the problem by removing exclusive feelings of guilt often suffered by the mothers. (Lewis B. Holmes, M.D., "Genetic Counseling for the Older Woman: New Data and Questions," *N.E.J.M.*, June 22, 1978.)

and three months later she was pregnant again. This time, amniocentesis revealed a normal number of chromosomes, and a baby girl.[24] At 43, she was delivered of a second child, a normal baby boy — again after a monitored pregnancy. With about 97 percent of all amniocenteses revealing *no* presence of the defect under suspicion, prenatal diagnosis provides incalculable reassurance to thousands under stress. Just as important, it opens the way to raise a family, when in the past, fear of the unknown would have been a powerful deterrent.

The list of disorders that can now, or will in the not-too-distant future, be diagnosed in utero is growing steadily. Not yet in routine clinical use but extremely promising are techniques for detection of sickle cell disease, thalassemia and muscular dystrophy. Every year, as many as 9,000 babies are born in the United States with one of the most serious of congenital malformations — a neural tube defect including spina bifida. It has been known for several years that when a fetus is affected, the amniotic fluid contains a large amount of the substance alpha-fetoprotein, and at a specific period during the pregnancy, it can also be detected in the mother's blood serum. The problem is that nine out of ten such babies are first-time occurrences in the family and there was no early warning system for them. Now, however, it has been established, both in England and in the United States, that routine screening of all pregnant women is feasible and accurate. Says Harvard University's Dr. Aubrey Milunsky, who recently completed a study involving 5,000 pregnant wom-

---

[24] Of the 23 pairs of chromosomes, 22 are somatic — the same in both sexes. The twenty-third pair is XY in males and XX in females. Learning the baby's sex in utero is one of the side benefits in amniocentesis.

en, "We are confident that maternal AFP screening, after FDA licensing of reagents, will eventually become part of routine obstetric practice."[25]

A fetal diagnosis — hailed as a first of its kind — was recently accomplished by a team of scientists that spanned the oceans. University of Texas Southwestern Medical School's Drs. Michael S. Brown and Joseph L. Goldstein[26] and associates at the University of Leuven in Belgium identified a fifteen-week-old fetus with a severe cholesterol problem.

It was the same hereditary disorder that had caused the woman's first child to die of a heart attack at the age of 8. The family history was extensive. The father has a mild form of hypercholesterolemia, and his father died at 47 of a heart attack. The mother has an elevated cholesterol, as do her two brothers, and *her* mother died of an unspecified heart condition at 54. Homozygous familial hypercholesterolemia is extremely rare, maybe one in a million, but for this family, the one in a million odds came up twice. Forewarned by the amniocentesis, they were spared the coronary death of a second preteen child.

The experience of the Belgian family is so rare, its counterpart may not be seen in this century. Coronary heart disease in families, however, is not rare, and recognizing those at risk does not require a tool like amniocentesis. The profile of the coronary-prone can now be drawn with a few simple tests and measurements.

Although the genetic defects in families with high blood pressure are still shrouded in mystery, discovery of the disorder involves no more discomfort on your part than rolling up your sleeve. If you are not the first in your family with diabetes, you do not have to know the precise

[25] *New England Journal of Medicine*, March 30, 1978.
[26] See Chapter 2.

pattern of inheritance to control it. And if your vision is threatened by cataracts or glaucoma, it can still be saved, regardless of your family history. As for longevity, the chances are overwhelming today that you will live longer than your parents.

The fact is that efforts to change your phenotype, regardless of your genotype, are becoming increasingly successful today. This is underscored by the waning of the epidemic of CHD, especially in the vulnerable middle-aged man, and the dramatic plummeting of the incidence of strokes and other complications of high blood pressure.

The key is in early recognition of your vulnerability and in adopting a lifestyle that can extend the vigor of youth and middle age all the way to the end of a maximal life span. The following chapters will tell you how.

2

Coronary Heart Disease — The Waning Epidemic

Two months after his heart attack, David Miller arrives at his scheduled checkup with a list of questions for the doctor. Looking trim and tan, and feeling fit, he wants to know, "When can I return to work? How soon can I reschedule the out-of-town business trip I was forced to cancel? What can I eat when I am away from home so that I do not come across as an invalid? What about sex?" All relevant and timely for a busy, successful 47-year-old man who has never been sick before.

His wife Sally has a few questions of her own, but her overriding concern by far is "How do we live now so that it never happens again?"

Before answering, the doctor makes a surprising suggestion of his own. In the light of what is now known about heredity and coronary heart disease, particularly in a man of David's age with his family history (his mother and her older brother both died of a heart attack in their early fifties), the next step should be a "look at the children."

George, at 16, is remarkably like his father — bright, ambitious and pushing to finish high school in three years so that he can start college early. The twins, at 14, are as different as sister and brother can be. Beth, mature for her age, is independent and sure of herself, while Benjy, the quietest member of the family, seems to be growing up in the shadow of the other two. With the anxiety of the last few months just beginning to recede into a bad memory, Sally questions the need to create a new anxiety for them now.

"After all," she pleads, "even with a family history, heart attacks rarely occur before middle age. What purpose would be served by involving the children while they're still in their teens?" Sally is right — a coronary attack *is* rare before the age of 40. But atherosclerosis,[1] the blood-vessel disorder that sets the stage for it, is not. Indeed, at the time of his mother's death, David, at 23, probably had already developed significant atherosclerosis that was to show up for the first time almost a quarter of a century later, the morning he woke inexplicably tired after a restless night's sleep.

"Bad dreams," he recalls. It was the first time in three years he missed the 7:31, the first time he complained of an almost unbearable crushing pain around his chest, and the first time Sally saw his usual robust color fade to a sweaty gray pallor.

The events that culminated in David's heart attack at 47

---

[1] A type of hardening of the arteries in which cholesterol and other materials in the blood are slowly and insidiously laid down as plaques in the walls of the blood vessels. The plaques can either close up an artery or predispose to formation of a clot that will block it. Tissues downstream from the obstructed artery, deprived of oxygen, will suffer serious damage. In the brain, this condition may cause a stroke; in the legs, severe pain in walking or even loss of limb; in the coronary arteries that nourish the heart, a heart attack.

began to operate many years earlier, possibly at birth. If he inherited his vulnerability from his mother, there is a 1 in 2 chance that he has transmitted it to each of his children. Doctors now know that with such a legacy, the future for George, Beth and Benjy may be not only "like father, like son," but also "like father, like daughter." If Beth has inherited her father's defective gene, her vulnerability may not be as severe as her brothers'. The danger of a heart attack may be delayed for as long as ten years, but it is there nevertheless. One study of twenty-one women under the age of 40 with advanced coronary atherosclerosis identifies a family history as the most important risk factor for them.[2]

Accounts of heart attacks date back to Biblical days. One of the earliest, in the Book of Samuel, is the case of Nabal, a man with a great appetite for food and drink, and a harsh temper. When Nabal's wife Abigail told her husband how she had resolved the serious problems he was experiencing with King David, his joy turned to rage and "his heart died within him and he became as a stone." Ten days later he was dead. Nabal's age is not given, but it is likely that he was in his prime, since he enjoyed such passionate devotion from a woman so desirable that King David made her his wife shortly after she was widowed.

More tangible evidence that atherosclerosis dates back to antiquity comes from an autopsy performed at Detroit's Wayne State University in February 1973. The body had traveled a long way in space and time. Name: Pum II; last known address: Philadelphia Art Museum; place of birth: Egypt; age at death: 35–40; age at autopsy: about 2100 years. Among the findings on the well-preserved mummy:

[2] J. Jurgen Engel, M.D., *et al.*, *Journal of the American Medical Association*, December 16, 1974.

"large and small atheromatous plaques in portions of the aorta . . ." and "thickening of the walls of a number of other arteries . . . typical of arteriosclerosis."

Today, while families with their inherited vulnerability, such as the Millers, have the greatest and most visible risk of early heart attack, they are only the tip of the iceberg. The genes for heart disease and other risk factors have been around for a very long time. Primitive man was subject to many stresses; high blood pressure has been around for centuries, and diabetes was described as early as 1500 B.C.[3] But only since the 1920s, following a radical change in lifestyle and eating habits after World War I, has coronary heart disease risen to epidemic proportions.

The diet of plenty includes too much fat (43 percent in contrast to the 30 percent of earlier years), too much refined sugar, and too many calories. Not only has overnutrition contributed to a sharp increase in diabetes and high blood pressure, but obesity has been reinforced by a sharp decrease in physical activity in everyday life. Apartment dwellers no longer climb stairs; suburban dwellers do not even leave their cars to open the garage door; golfers ride carts as often as they walk. The automobile makes an additional contribution to coronary heart disease with the carbon monoxide it injects into the air. And it is only since the 1920s that cigarette smoking has become widespread.

David Miller is one of 1¼ million Americans who last year became a victim of what the World Health Organization describes as the "most serious epidemic facing mankind today," killing over half the males and a large percentage of the females in the western industrialized world and now emerging as a threat in the developing countries as well. The twentieth-century epidemic of coronary heart

---

[3] Both powerful contributors to coronary heart disease. See Chapters 3 and 4.

disease has a number of striking features. It has a predilection for middle-aged man.[4] It flourishes in an environment of economic growth and affluence. It has an unprecedented period of incubation. If you are exposed to measles and are susceptible, you will break out in a rash in a week or ten days, but with atherosclerosis (to which all Americans are susceptible), it may be twenty or thirty years or more before the first symptoms appear. Unlike other epidemics, it has varied and complex causes — nutrition, behavior, emotion and heredity. In fact, it is associated with one of the most common human genetic defects.

Whether the underlying atherosclerosis is triggered by a defective gene or is acquired through predisposing lifestyle and experiences, the end result is the same — premature coronary disease. Among American soldiers in their early twenties killed in the Korean War, the majority of those autopsied already showed evidence of atherosclerosis. Sixteen years later, young casualties in Vietnam showed the same damage to their blood vessels. That this is not a phenomenon of fighting men is underscored in a New Orleans study of a series of 15- to 19-year-old boys and girls who died in accidents. They, too, had changes in their coronary arteries. Had they survived the wars and the highways, many might well have been among the 648,540 Americans who died last year from heart attacks, with men outnumbering women 3 to 1. More than 200,000 were between 40 and 59 and, like David, at the peak of their productive lives. For many, the first symptom

---

[4] While the risk of coronary heart disease increases with age, so great is its rise in the middle-aged man that life expectancy, which has risen twenty years in the United States since the turn of the century, applies only to children. For the fifty-year-old man, the increase until recently was only a little more than two years, reflecting the toll of coronary heart disease. A new (July 1977) projection now promises an extended life span of three years.

was the last — sudden death within hours of the attack.

Nevertheless, despite the gloomy statistics, there is a new optimism building. After raging for forty years, the epidemic appears to be waning; there has been an impressive and continuing decline in coronary heart disease deaths in the last decade, notably in the middle-aged man. To be sure, some of the new success is due to better medical management and more sophisticated hospital care for the heart attack victim. But "much of the credit," Dr. Robert I. Levy, Director of the National Heart, Lung and Blood Institute of the National Institutes of Health, told a July 13, 1977 press briefing,[5] "belongs to the voluntary changes in diet adopted by the American people after a decade of publicity on the dangers of overeating high fat and high cholesterol foods."[6]

There is a new optimism in the response to many of the 4 million heart attack victims and their spouses when they ask, "How do we live now so that it does not happen again?" Until recently, the routine advice to "eat less, exercise more, stop smoking, keep your blood pressure down, and stop worrying" was a framework in which to function. Not so certain was the promise that heeding the advice would actually unclog the clogged arteries. Today, for the first time in history, researchers no longer view atherosclerosis as a hopelessly degenerative disease. Based on a growing body of evidence from studies on man and animals, they report with confidence that it may well be ". . . almost completely preventable and . . . substantially reversible."[7]

Perhaps the most heartening news is that the focus is

[5] Subject was preliminary findings of six-year-old Lipid Research Clinics program.

[6] Particularly among better educated, professional and managerial persons.

[7] University of Chicago Drs. Robert W. Wissler and Dragoslava Ves-

slowly but surely shifting to the child and adolescent — when atherosclerosis starts — and to the physician who cares for him. Says Dr. Ernst L. Wynder, Founder and President of the American Health Foundation, "Pediatricians have, without doubt, done exceptionally well in keeping our children healthy, but at the same time they have fallen far short of creating healthy adults."

Today, the momentum toward that goal is beginning to take shape, underscored most recently at the October 1978 First International Symposium on Primary Prevention in Childhood of Atherosclerotic and Hypertensive Diseases. Presented by the Chicago Heart Association, the American Heart Association, and the American Academy of Pediatrics, and co-sponsored by the NHLBI and the World Health Organization, it brought together many of the world's leading researchers in the field. Also indicative of the need to bring the pediatrician into this new area of preventative medicine was the International Symposium in Pediatrics, presented in September 1978 by the University of Montreal Faculty of Medicine.

One of the earliest alerts came from Dr. Levy in 1972 when he told the American Academy of Pediatrics, "The pediatrician must become a student of atherosclerosis." The following year, renowned cardiologist Dr. Paul Dudley White called for a "children's crusade" as the most logical step in fighting coronary heart disease. In 1976, the NIH Task Force on Genetic Factors in Atherosclerotic Disease, which read like a roster of Who's Who in the field, urged that one of the earliest priorities of physicians concerned about coronary heart disease should be a hard look at the children of coronary heart disease victims who were struck at an early age. By 1977, the role of heredity was a recurrent theme at the Science Writers' Forum in San

selinovitch, *Modern Concepts of Cardiovascular Disease* (American Heart Association), June 1977.

Antonio, where the American Heart Association brought together more than two dozen of the leading researchers in the country to discuss their most recent work, both clinical and investigative.

It has been known for many years that when early coronary heart disease runs in families, more than a common environment is involved. Identical twins are more likely to be affected than brothers and sisters. But brothers and sisters share a greater risk through their genetic endowment than husbands and wives do through their shared environment. During the forties, fifties, and sixties, mounting evidence was uncovered to implicate heredity, but it was not until the early 1970s that a landmark Seattle study turned up three distinct genetic defects in fat metabolism among 500 survivors of coronary attacks and 2,500 of their relatives.

Acclaimed as the first clear-cut genetic clue to the mystery of coronary heart disease, the big surprise was not the family pattern — it had long been assumed that a group of genes acting together and interacting with the environment was the culprit (and that is still true for large numbers of victims). But the discovery that at fault in each of these three disorders is a single gene was unexpected. With one in every 150 persons in the United States carrying one of these genes, familial hyperlipidemia (increased fats in the blood), *inherited vulnerability to early coronary heart disease, may well be the most common human single-gene disorder*, more widespread than sickle cell, Tay Sachs, hemophilia and cystic fibrosis.

Among the Seattle survivors under 60, 20 percent had a significantly elevated level of blood fats, due to a genetic defect: (1) cholesterol, the ubiquitous fat-like substance in your body, essential to many life processes, but a sign of grave danger when too much piles up in your blood —

indeed, the single best predictor of risk for coronary heart disease; (2) triglycerides, the form in which fats are stored in fat tissue, to be drawn on for energy, etc., when needed; (3) a combination of both, which unexpectedly turned out to be more common than either the cholesterol or the triglycerides alone.

Inheritance of the gene for hypercholesterolemia shows up at birth — in umbilical (cord) blood, in fact. If present, it can easily be identified in the Miller children now. Elevation in triglycerides is usually not manifest until adolescence or early adulthood, but, in any event, long before serious coronary heart disease develops, and in ample time to take preventive measures. "The anxiety produced by the statistical prediction of heart disease," says Task Force member Dr. Arno Motulsky, Chief of Medical Genetics, University of Washington School of Medicine, "should be outweighed by the certainty that existing therapeutic measures will defer the onset of coronary heart disease."

When the families of the Seattle survivors were studied, the picture that emerged was that of a dominant autosomal[8] disorder — where only one dose from either parent is required to transmit the vulnerability, and sons and daughters are at equal risk. While many families with familial hypercholesterolemia had been described, without a genetic marker, confirmation of this pattern of inheritance relied on good family pedigree studies — preferably a very large family, spanning three or four generations, with the majority still alive, and good records on the deceased members.

Just such a family was brought to the attention of the University of Washington team[9] conducting the study.

[8] See page 16, Chapter 1.
[9] Drs. Joseph L. Goldstein, William R. Hazzard, Helmut C. Schrott, Edwin L. Bierman, and Arno G. Motulsky.

Surprisingly, this family was from a part of the world and a society where coronary heart disease is a rarity — it was an Eskimo family from a small Aleutian village.

The first member of the family the researchers were to meet was a 22-year-old woman with all the signs of familial hypercholesterolemia — blood cholesterol three times normal (600 milligrams), small fatty deposits on the skin (xanthoma), and a halo[10] in her eyes — but no sign of heart disease yet. She was one of ninety-two descendants of a couple who had settled in the village more than sixty years earlier. They had eighteen children, sixty grandchildren (of whom the young woman was one), and fourteen great-grandchildren. The youngest member of the family was less than a year old, and the oldest, the great-grandfather and the original settler, was alive and well at 87. Virtually all living members of the four generations were still in the village.

When they were tested, the outcome was almost a classic example of how a dominant autosomal gene is transmitted. About half, with a normal cholesterol, had not inherited the defect, while the half that had inherited it (heterozygote) had a mean cholesterol of double normal. Not only were the latter themselves candidates for coronary heart disease, but each could transmit it to each of his or her children. The one in a million persons who inherits a double dose of the defective gene (homozygote), one from each parent, runs an extraordinary risk of early heart disease, rarely surviving beyond the age of 20.[11] There were no homozygotes in the Aleutian family, and no child of two normal parents had elevated cholesterol; but the

[10] Arcus cornea — a deposit of fatty granules in the form of an opaque grayish ring at the margin of the cornea and sclera.

[11] See Chapter 1, page 21, for the story of the first prenatal identification of the one-in-a-million who has inherited the defective gene from each parent.

affected children with one affected parent showed the signs early in life. In addition to the increased cholesterol, they also had xanthoma or a halo in the eye or both. "Most important," reported Dr. Goldstein, "50 percent of the adults with elevated cholesterol had documented evidence of coronary heart disease," and without intervention, each was a candidate for a heart attack between the ages of 40 and 60.

Where did this relentlessly persistent gene come from in an Aleutian family? From the great-grandmother, who died of a heart attack at 62. She was not all Aleutian — her ancestry was part Aleutian and part Russian. Most likely it was from her European ancestor that she inherited the gene.

Another milestone in genetics was achieved within a few short years after Dr. Goldstein moved on to the University of Texas Southwestern Medical School, where he now heads the Division of Medical Genetics. He soon began a collaboration with another young researcher, Dr. Michael Brown, and in 1974 the team disclosed that not only had they identified the nature of the genetic defect (an extremely important advance in understanding a dominant disease), but also that they had found what Dr. Goldstein describes as "an unexpected dividend . . . the first good handle on the biochemical mechanism by which cholesterol is laid down." [12]

On the surface, the Aleutian family and the Miller family have little in common, but they do share the same inherited vulnerability to coronary heart disease. David's cholesterol at the time of his attack was high — over 350 milligrams. By this time, Sally has been convinced of the wisdom of testing the children. "The attack to prevent is the first," the doctor advises, "and the time to start is now."

[12] For these accomplishments, Drs. Goldstein and Brown have been honored with prizes both in the United States and abroad.

Of the three, both boys are within normal limits. Beth, however, has a cholesterol far above normal limits for a 14-year-old, in the absence of other disorders that drive cholesterol up.[13] When the early warning system was extended to David's brother Milton, nursing a tennis elbow but otherwise in good health, he was found to be at no greater risk than any other 49-year-old American male. Their younger sister, Rosalie, on the other hand, a grandmother at 45, had inherited the family gene for hypercholesterolemia. Fortunately, she had not transmitted it to either of her daughters, and because a dominant trait does not skip a generation and then show up again, her daughter's 2-year-old was not at risk.

As in many instances of genetic counseling, the pedigree study in the Miller family turned up fewer members at risk than not, and provided more assurance than anxiety. For Beth, whose lifestyle is not yet set, and her Aunt Rosalie, however, it has far-reaching implications for their future well-being. Some of the still-unexplained advantage[14] that premenopausal women have in staving off early heart attacks is lost to them. It is not too late to take steps to protect Rosalie, and not too early to intervene for Beth.

When the time comes to make a choice of contraceptives, Beth will be alerted to the increased hazard of the Pill, for estrogen raises blood fats as well as intensifying other coronary heart disease factors. Aunt Rosalie went off the Pill at 40 — even before she knew about her special

[13] Thyroid disease, certain types of kidney disorder, liver disease, and *too large a fat intake in the diet.*

[14] Some of the theories about the source of the natural superiority of women involve hormone activity, enzymes that control hormones, elimination of potentially atherogenic substances through the menstrual fluid, and a higher blood level of one type of cholesterol that protects against coronary heart disease. ". . . obliterating this female advantage are diabetes, *familial hypercholesterolemia* [Authors' italics], advanced age and menopause." (William B. Kannel, M.D., *J.A.M.A.*, June 6, 1977.)

risk. Indeed, given the widespread incidence of atherosclerosis in teenagers who do *not* inherit the defective gene, the alert for the future is just as appropriate for George and Benjy as it is for Beth.

A reminder to the profession that "atherosclerosis starts in adolescence" comes from Dr. William Kannel, Director of the Framingham Study.[15] "Rarely does the physician have such an opportunity to detect potential victims years before onset of symptoms . . . and take steps in time to prevent or at least delay not only illness but even death." And rarely, it might be added, have potential victims had the opportunity to protect themselves from a threatened chronic illness in their future.

Prevention became a possibility with two events in the mid-1960s. First was the profile of the coronary-prone individual that emerged from the Framingham studies, which identified risk factors. Most encouraging was the knowledge that the big three — elevated serum cholesterol (the single best predictor in a man under 60), high blood pressure, and cigarette smoking[16] — can be successfully modified or eliminated. Also contributing are abnormal sugar metabolism and diabetes, gout, obesity, abnormal electrocardiogram, personality, and family history. A combination of *all* the risk factors adds up to a risk thirty times that of the individual who has none. For a man in his forties, the risk is fortyfold.

How valid is the profile? In 1973, the American Heart

---

[15] Taking its name from the New England town where a coronary-prone profile was drawn over twenty years of observation of 5,209 men and women in a study undertaken by the United States Public Health Service in 1948.

[16] Eighty percent of disability and death from premature atherosclerosis and coronary heart disease occurs among persons who have one or more of these three risk factors. One doubles the risk, and all three impose a tenfold risk!

Association released its *Coronary Risk Handbook*, designed to provide doctors with a relatively simple tool to estimate risk of coronary heart disease in the next six years in patients with *no* clinical evidence, focusing attention on the young and vulnerable, when intervention will be most successful. Based on the Framingham studies and tested on persons of known risk in Chicago, Los Angeles, Tecumseh, Minneapolis and Albany, its accuracy has since been confirmed in a June 1977 report in the *New England Journal of Medicine*. When Dr. Atone F. Salel and his colleagues at the School of Medicine of the University of California at Davis studied a series of patients undergoing cardiac catheterization, they found that the risk profile not only predicted the disease, but also correlated significantly with the severity of damage to the coronary arteries.

The second preventive event of the sixties was the recommendation in 1964 by the American Heart Association that a change in the American diet (limiting saturated fats and cholesterol) could help bring about a reduction in heart attacks and strokes. The message has come through to a significant number of men in their forties and fifties and their wives. "Probably more Americans are concerned about an increase in their cholesterol than about antiballistic missiles," Dr. Donald F. Fredrickson, now Director of the National Institutes of Health, told the Royal College of Physicians of London in 1969.

By 1975, the market baskets of Americans contained 32 percent less butter than during the previous decade, 12 percent fewer eggs, 56 percent less animal fats, and 44 percent more vegetable fats and oils.[17] By 1977 they were not only asking, "Is my cholesterol down?" (it has dropped 5 to 10 percent below levels of two decades ago); their

[17] United States Department of Agriculture.

questions reflected new insights and a new vocabulary — LDL, VLDL (the cholesterol packets in their blood that spell danger) and HDL (the cholesterol packet that promises protection).

Over a ten-year period, smoking has declined among men by 14 percent, and among women by only 3.9 percent. Both groups are losing weight (unfortunately, not all keep it off), and becoming more responsive to their doctor's admonitions when told of elevated blood pressure. Yet throughout this period, concern for their teenagers, who might already be handicapped with two or three risk factors, lagged. Teenage smoking is increasing relentlessly, a typical teenage diet is guaranteed to push up cholesterol, and increased blood pressure and obesity are not rare among teenagers.[18]

Today, however, parents of school-age children, long concerned about IQ, achievement tests, grade-point averages, SAT, etc., are adding to the profile the numbers for cholesterol and blood pressure. If you are a parent in the midwestern town of Muscatine, Iowa, the first inkling that your family may be at risk for premature coronary disease may have come from an NHLBI study in which your child is participating. On the other hand, you may be all too familiar with the toll coronary heart disease has taken among your close relatives but still be surprised that the same vulnerability is already showing up in your 10-year-old.

Long recognized for the pearl buttons it produces, Muscatine, Iowa, is now writing a chapter in public health history as the site of the first five-year follow-up study of

[18] At the July 13, 1977 National Institutes of Health briefing, Dr. Basil M. Rifkind, Chief of the Lipid Metabolism branch, noted a drop in adolescent cholesterol levels but acknowledged that the reasons are not yet completely understood.

coronary heart disease risk factors in school-age children. Almost 8,000 students, ages 5 to 18, were screened for cholesterol, triglyceride, blood pressure and obesity. Three surveys (1971, 1973, 1975), with a total of about 2,000 participating, in all the surveys, revealed dismayingly high values in all areas. "About 5 percent of the entire population of school children," reported University of Iowa's Dr. Ronald M. Lauer to the 1977 Science Writers' Forum, "have a cholesterol level over 230 milligrams — a risk figure for adults. On repeated measurements, 1.5 percent remained high." And a look at the families of that 1.5 percent revealed that they had seven times more first-degree relatives (parents, aunts or uncles) who died of coronary heart disease than other participants did.

"We believe it is important for physicians who provide medical care for school-age children to routinely measure serum cholesterol," Dr. Lauer stresses. "If it is high, it should be measured several times, and if the high level persists, this is justification for the physician to intervene with the entire family, giving detailed advice on the need for a low-fat, low-cholesterol diet." How do families accept the advice? "Very well," says Dr. Lauer. "After all, some of the families have already had bad experiences among their relatives, some while still in their thirties."

The first children's crusade, in which the children are running their own detection and prevention program, involves 3,000 11- to 14-year-olds in New York City and Westchester. Known as KYB (Know Your Body), it was conceived and is directed by the American Health Foundation, and includes a search for familiar risk factors for heart disease, cancer and stroke — weight, blood pressure, smoking and cholesterol. To complete the picture, they fill out a family history (assembled at home with the family) and record their eating habits and preferences.

In an interview, Program Director Dr. Christine L.

Williams, whose expertise is in both pediatrics and public health, spoke of some of the findings of the first year of the three-year program. Almost 20 percent have cholesterol too high for their age group, 12 percent are overweight, 10 percent are smoking (more girls than boys), and 20 percent scored only "fair to poor" on an exercise test.

As in Muscatine, the American Health Foundation includes in its program informing parents of the results with the appropriate follow-up advice — see your doctor, change your diet, etc. The KYB program in New York goes further. In addition to designing a special health education program, including a Teacher's Guide, program directors initiated a special school program geared to helping children at risk for disease because of overweight or high cholesterol achieve and maintain an ideal weight through better nutrition. The program is supervised by the school nurse or a faculty member.

More than 70 percent of the students said "yes" when asked to participate in the KYB program (with parents' consent a prerequisite). Some were encouraged to sign up when they saw their teachers volunteer. They appreciate seeing their own results, which they record into a personal "health passport," and they are beginning to accept responsibility for their own good health. In contrast to many doctors who say "don't worry" when a potentially threatening number turns up in a child, Dr. Williams says, "We *want* children to be concerned — indeed, concerned enough to act." And enlisting children and teenagers in the battle may be the essential ingredient required for a successful crusade.

Will tampering with the diet of a growing teenager interfere with his development? "It is an absurd assumption that teenage diets are perfect," says Dr. Robert B. McGandy, Harvard School of Public Health, citing a four-year study he conducted on eating habits of more

than 1,200 teenagers. "Their diets are already more con-
ducive to atherosclerosis than that of the typical adult
American male, and their rising serum cholesterol reflects
it." Surveys in Vermont and Iowa turned up elevated
cholesterol levels in 13 to 15 percent of the teenagers, and
in Wisconsin, a staggering one in three children had
cholesterols over 200 milligrams when screened. For most,
it is diet-induced and can be brought back to normal
without radical changes. Cutting down (calories, animal
fats, sugar, etc.) and not cutting out is the formula.

Teenagers with familial hypercholesterolemia, such as
Beth, whose cholesterol has been piling up since birth,
will need more intensive intervention and a lifelong adherence to a diet. Some older children may be treated with a
cholesterol-lowering drug if diet alone fails. The younger
the child, the greater are the odds that the response to a
change in diet will be successful. At the University of
Cincinnati Medical Center, Dr. Charles J. Glueck and his
colleagues succeeded in maintaining essentially normal
cholesterol levels in infants identified from cord blood,
placed on a diet low in cholesterol and high in polyunsaturates, and tested at six months, a year, and eighteen
months.[19] When given a "normal diet" at eighteen months,
the cholesterol went up! Another group of 6- to 11-year-
olds, when allowed to eat what they wanted, registered
"marked elevation in cholesterol — in the range of 320 to
380 milligrams. Substitution of a diet low in cholesterol
and high in polyunsaturated fat brought a prompt drop
toward normal. *Older children with comparable initial cholesterol levels did not respond as dramatically to dietary modification.*"

What does it mean to a young child to be on such a diet?

---

[19] There has been some concern, based on animal experiments, that such
a diet might be detrimental to normal growth and development.
Evidence today suggests that this is not so.

Adam Jr. does not think he is odd. He is not the only 10-year-old in the neighborhood who has been drinking skim milk for the last few years, nor does he miss the butter cookies, ice cream and frequent hamburgers his friends consume so freely. Moreover, he adores his father, Adam Sr., who eats pretty much the same menu. "Even without the doctor's insistence," his mother comments, "he would find it a great way of life, with his love of sports and his goal to be fit."

So it has been since shortly before his sixth birthday, when he and 4-year-old Suzy were both tested following Adam Sr.'s heart attack at 37. Adam's cholesterol was abnormally high for a 6-year-old — as high as that of the average 40-year-old. Suzy's was normal. Four generations in Adam's family is another striking example of how genetics works and the role chance plays in whether the normal or defective gene is passed on to each offspring in each generation.

"I've known since I was 22 that I had a high cholesterol, the year my father died of a heart attack at 49," Adam Sr. recalls, "and I knew then that I should be careful, but no one really spelled out for me what 'being careful' meant. Anyway, I figured that with all the new discoveries, the doctors would have something in the way of a cure by the time I was in my forties."

What Adam Sr. did *not* know was that he could have been the one in a million to inherit the defective gene in a double dose, for, through a strange twist of fate, his mother too carries the defective gene. Her cholesterol was over 400 milligrams when she suffered a serious stroke at age 45 (she is fine now). *Her* father died suddenly at 57, with no warning of any illness; one brother died of a heart attack at 41; and another is making a good recovery from a coronary attack at 55. If each of Adam Sr.'s parents had transmitted the good gene, he would have been at no greater risk than the average American male. If each had transmitted the defective gene, he would have inherited the most virulent form of familial hypercholesterolemia, whose victims rarely live beyond 20. But one transmitted the good, the other the bad. At this point, no one knows from which parent the legacy came. When his own children were conceived, again it was purely by chance that Adam Jr. got the bad gene and Suzy the normal one. Today, with the new knowledge about atherosclerosis, Adam Jr. can be reassured with more than just an admonition to be careful. Despite his legacy, he is the first victim in four generations with the hope of a long life.

As far back as 1896, the world-famous Dr. William Osler, who viewed atherosclerosis as "the Nemesis through which Nature exacts retribution for the violation of its laws," focused on change of diet as the keystone of both prevention and treatment. More than forty years were to pass before the specifics of a non-atherogenic diet and its power to protect from coronary heart disease were spelled out by a renowned Dutch physician based in China in the late 1930s. Dr. I. Snapper was Professor and Head of the Department of Medicine, Peiping Union Medical College, Peiping (now Peking). He had been invited to Peiping to teach, but he learned there as well, and published what he learned in his 1941 *Chinese Lessons to Western Medicine*.

Commenting on the virtual absence of coronary heart disease in North China (three possible cases in two years), he wrote that "the rarity is the more striking because the frequency . . . in America and Europe is appalling." Just as striking to Dr. Snapper was the difference in diet, with the Chinese diet very low in cholesterol, high in unsaturated fats, and virtually devoid of butter and milk. A typical day's menu at that time included bread made of a corn-soybean mixture and vegetables which varied with the seasons — string beans, spinach, eggplant, turnips, kohlrabi — all sauteed in sesame oil. Eggs were an infrequent luxury, and meat was eaten by the poor only on New Years and on the Tenth Day of the Tenth Month. Total calorie intake was low. Not surprisingly, the average blood cholesterol was as low as 100 miligrams (in contrast to the 230 to 250 of the average "normal" American male today). But not all Chinese at that time were immune to coronary heart disease. Rich Chinese in what was then the Dutch East Indies, who adopted the rich western diet, developed the disease at as high a rate as the Europeans.

"We are digging our graves with our teeth," Dr. Snapper told us in the early 1960s, as he deplored what he considered our intransigence in not learning the Chinese lesson.[20] Included in the 1941 lesson was also the observation that "smoking of cigarettes is relatively rare here."

Dr. Snapper's China observations have since been confirmed on a global scale.[21] In countries where the mean cholesterol runs high, including the United States, Britain, Finland and Norway, coronary heart disease is rampant. Where the cholesterols are low, as in Japan, coronary heart disease is low. The primacy of cholesterol is highlighted by the situation in Japan today. Despite widespread high blood pressure and excessive cigarette smoking — both powerful risk factors — the coronary heart disease rate is very low, reflecting the low cholesterol in the population. Indeed, the upper limit in Japan is not too different from the lower limit in Britain.

On the other hand, when Japanese move to California, their coronary heart disease rate rises tenfold, and so it is

[20] Interned by the Japanese during World War II, Dr. Snapper was released to the United States in exchange for a high-ranking Japanese national being held in the United States. He became an American citizen and continued to teach and treat until his death in 1973.

[21] Ancel Keys, "Coronary Heart Disease: The Global Picture," *Atherosclerosis*, 1975.

for the Yemenites who move to Israel, the Irish who move to Boston, and the Neapolitans who move to New York. Moreover, there is a "Changing Prevalence of Coronary Artery Disease in People's Republic of China" today.[22] The incidence of heart attacks among Chousan fishermen over 40 is still a low 0.6 percent, in contrast to 7 percent among Shanghai factory workers. Wherever the cholesterol levels are high, the diet is rich in animal (saturated) fats — beef, butter, eggs, whole milk, etc.

In 1954, Rockefeller University's Dr. Edward Ahrens, Jr., showed in his experimental work that it is possible to lower cholesterol levels by substituting certain vegetable fats (polyunsaturates) for the animal sources, and a new era in dealing with coronary heart disease was launched.

Despite the indisputable association of cholesterol with atherosclerosis and heart disease, you probably know at least one person who eats eggs every day and ice cream every night, and seems to get away with it. On the other hand, there are others on a relatively prudent diet who, nevertheless, have high blood cholesterol and suffer from early coronary disease. The fact is that even within populations where coronary heart disease and cholesterols are uniformly high or uniformly low, there is a wide range of individual differences. Clues to the puzzle come from more than ten years of research at the Bowman Gray School of Medicine, Wake Forest University. Among pigeons raised in identical situations and fed identical diets, some had cholesterol levels of 300 milligrams, while others had levels in excess of 2,000 milligrams. By selective breeding, the scientists succeeded in developing strains of "low responders" and "high responders" and have recently nailed down the biochemical difference between them — the activity of an enzyme[23] that converts excess cholesterol

---

[22] *Annals of Internal Medicine*, 1974.
[23] Cholesterol-7-alpha-hydroxylase.

to a form in which it can be excreted. The low responders develop minimal atherosclerosis, while the high responders, loaded with cholesterol, develop extensive atherosclerosis.

Studies with a closer relative to man — the South American squirrel monkey — provide additional data on the role of heredity in cholesterol metabolism and atherosclerosis. "About 65 percent of the variability in the pigeons and the monkeys," says Dr. Thomas B. Clarkson, "is attributable to genetic factors." Furthermore, the new look at prevention and treatment of atherosclerosis, which goes beyond cholesterol as the sole culprit and into the blood vessel in which it will do its damage, promises — for the first time since the cholesterol connection was first made — to dispel the mystery that still surrounds coronary heart disease.

Cholesterol's contribution to atherosclerosis was first demonstrated early in the century by the Russian scientist N. N. Anitschkow, who produced the disorder in rabbits (immune by nature) by adding egg yolks and cholesterol to their diet. He also drove up their level of blood cholesterol and other fats. A tasteless, odorless, colorless, fat-like substance, cholesterol is the Dr. Jekyll and Mr. Hyde in your body, for it is essential to life. Indeed, it participates in so many processes that most, if not every, cell of your body can make it from simple substances, although most of it is synthesized in the liver. It is a building block for cell membranes, a precursor for bile acids and a variety of hormones. One form of cholesterol, it is now known, *protects* against heart attack — but more about that later. You have a complex and effective feedback mechanism that ensures that the liver will supply what you need, and that slows synthesis down when you ingest more than you need. "Most Americans," says Dr. Goldstein, "take in at least four times as much cholesterol as they need to maintain good health." The ability to handle the excess is the difference between good health and life-threatening disease.

The foundation for today's knowledge of the role of increased cholesterol and other blood fats in heart disease was laid in the late 1940s and early 1950s by a young research nuclear chemist and physician, Dr. John Gofman, who at the time was Professor of Medical Physics at the University of California at Berkeley. After wartime service with the Manhattan Project, where he was co-discoverer of uranium 233 with Nobel Laureate Glenn Seaborg, Dr. Gofman was one of the new breed of scientists dedicated to applying sophisticated know-how from the fields of physics and chemistry to problems in biology and medicine.

Although it was known by that time that increased blood fat predisposed one to coronary heart disease and that there were both familial and environmental factors at work, the total fats were not revealing enough. Because fats, including cholesterol, do not mix with the watery medium of blood, they are carried in the bloodstream in packets, piggyback on proteins. The combinations are called lipoproteins. Each lipoprotein packet contains cholesterol, triglyceride, phospholipid and protein, each with different proportions of the fats. The total amount of cholesterol — the number you get in a routine blood test — is the sum of *all* the cholesterol in all the different lipoproteins. That value is a good predictor of vulnerability to heart disease in men and women under 50, but beyond that age it is not revealing enough.

Dr. Gofman discovered that, at any age, how the respective fats are carried is more important in predicting risk for atherosclerosis and heart disease than the total sum. By spinning blood samples in an ultracentrifuge at 60,000 revolutions per minute, he separated out nine different packets in the transportation system. One fraction that floated to the top because it was of low density (LDL) was rich in cholesterol. Another fraction, which was even lighter and of very low density (VLDL), was rich in triglycerides. *Both were associated with a high risk for coronary heart disease.* At the time, he also described a high density factor (HDL), and he noted that the higher the LDL and VLDL, the lower the HDL.

Initial reaction to Dr. Gofman's work was mixed. Some investigators in the field were not too impressed, while others, excited about the new approach, were spurred to apply it in their own investigations.[24] In the 1960s, Dr. Fredrickson, Dr. Levy and an NIH colleague, Dr. Robert J. Lees, brought lipoproteins out of the research lab and into clinical practice. Using a simpler technique (electrophoresis), they studied lipoproteins in 400 coronary heart disease victims and their families and identified five abnormal patterns of fat transport. They found that two of the five — Type 2, which is high in cholesterol (LDL), and Type 4, which is high in triglycerides (VLDL) — carry a big risk for inherited heart disease.

Among the Type 2 families are those with the genetic defect discovered by Drs. Goldstein and Brown, which accounts for the hypercholesterolemia in the Aleutian family, the Miller family, and Adam Jr.'s family. The rate of cholesterol production is normally controlled by an enzyme within the cell (HMG CoA reductase), which steps up its activity when more cholesterol is needed and stops producing when it gets the message that the level of cholesterol in the blood is adequate. In familial hypercholesterolemia, the message does *not* get into the cell be-

---

[24] One of us (I.J.G.) was one of Dr. Gofman's early pupils in learning the complex technique of lipoprotein analysis. Upon my return to what was then called Beth-El Hospital, in Brooklyn, New York, in 1952, I began my first studies in lipoproteins and coronary heart disease.

cause of a deficiency of LDL receptors on the surface of the cell that normally permit the cholesterol to enter. With communication disrupted, the enzyme is not turned off and the synthesis of cholesterol continues at an inexorable rate. More than a casual decrease in cholesterol intake is now required to keep cholesterol levels in check. Normally, each cell has about 250,000 LDL receptors on its surface. Cells of heterozygotes (individuals with a defective gene from one parent), such as David Miller, the two Adams, and the young Aleutian woman, have only half the receptors. The homozygote, with two defective genes, has virtually no receptors. The Dallas team is now looking for (1) a simplified test for receptors so that it will no longer be necessary to depend on family pedigree studies exclusively; (2) a substance that can fool the cell into getting the message to the enzyme by bypassing the inadequate receptors.

By this time, based on their studies, Drs. Fredrickson and Levy had already developed an extremely effective guide to lowering dangerous levels of blood fats — whether acquired by heredity or by lifestyle. Using "diet as the keystone," they put together a series of booklets for both physicians and patients, each booklet listing specific diets for each of the five lipoprotein disorders.[25]

Meanwhile, researchers were beginning to take another look at HDL — the lipoprotein fraction low in cholesterol and high in density — which Dr. Gofman had described in his earliest studies. Cornell University's Dr. David Barr suggested as far back as 1951 that HDL could protect against atherosclerosis instead of promoting it, but it is only in the last ten years that evidence has been pointing in that direction. When Dr. William P. Castelli[26] told the

[25] A handbook for your physician and a specific diet for you are available from the Office of Information, National Heart, Lung and Blood Institute, National Institutes of Health, Bethesda, Maryland 20014.

[26] Director of Laboratories, Framingham Study.

Science Writers' Forum, "Some of the cholesterol in your blood is associated with good health instead of bad health," he made headlines in both daily newspapers and professional journals. Indeed, following a report by special news editor Nathan Horowitz in *Medical Tribune* (the physician's newspaper), a record number of inquiries were received from doctors. "The response has been unbelievable, overwhelming," Dr. Castelli said. After his return to Boston from the San Antonio seminar, he was besieged by "hundreds of phone calls . . . stacks of mail . . . FANTASTIC."

How does HDL confer its benefits — and how can you raise your level to share in the benefits? HDL appears to protect by sweeping out excess LDL from the blood before it can be incorporated into a plaque, and carrying it to the liver for elimination. It also interferes with the basic process of atherosclerosis by blocking the uptake of LDL into the cells.

One way to have a high HDL is to be born into a family where your parents, grandparents and great-grandparents have high HDL and low LDL and to maintain that lipoprotein relationship throughout your life. Dr. Charles J. Glueck and his colleagues at the University of Cincinnati Medical Center studied eighteen families with the combination. The men in these families had a life expectancy of 77 years (compared to the average of 71), and the women could expect to live to 82 (compared to the average of 75).[27] The incidence of myocardial infarction among first-degree relatives (children, sisters, brothers, parents) was considerably lower than in a normal control population. What they all shared was a one-to-one ratio of LDL to HDL. By contrast, the "normal" control group showed a ratio of 2.5 LDL to 1 HDL.

Some ethnic groups have an HDL advantage. In rural

[27] *Journal of Laboratory and Clinical Medicine*, December 1976.

Greenland, Eskimos, whose total cholesterol (including the LDL) is high, have an even higher HDL. The incidence of coronary heart disease among the Eskimos is lower than among Danish males on the mainland.

It also helps to be a woman. The average American male has an average HDL of 45, while the average for women is 55. Dr. Castelli suggests that this may be one reason that coronary heart disease occurs less often among women. It may be that your HDL and not your sex predicts your coronary future. "At any level of HDL," says Dr. Castelli, "men and women show quite similar rates of coronary heart disease." The higher the LDL and the lower the HDL, the greater the risk for coronary heart disease.

Long-distance runners, particularly marathon runners, have high HDLs. Stanford University's Dr. Peter Wood, himself a marathon runner, reports average levels of 66, and his findings are borne out by studies of runners in New Orleans. In fact, Dr. Castelli, who is in his early fifties, well proportioned, dynamic and prematurely gray, now jogs.

Perhaps the best way to achieve a high HDL is not to lose the proportion with which you are born. You are born with half of your cholesterol in the form of HDL (an outstanding amount). On the typical American diet, that level declines as the more harmful factors — LDL and VLDL — go up with age. Can diet later in life turn that around? The Framingham researchers studied sixteen Zen communes from the East Boston hypertension survey. "They have low blood pressure, low total cholesterol, and those who added fish to their vegetarian diet had the highest HDLs," observed Dr. Castelli. Finally, a number of cholesterol-lowering drugs, including clofibrate and nicotinic acid, also appear to raise HDL.

Stressing the importance of knowing the HDL value in

a person whose total cholesterol is in the middle to high range (250–290), he spoke of a woman who was being treated for high cholesterol until a test disclosed an HDL of 120, "the highest we had found so far." Dr. Castelli predicts, "She will outlive her doctors."

Much of this new knowledge has come out of the Framingham Study since 1968, the year the history-making project was scheduled to come to an end. For twenty years, the 5,000 participants had remained remarkably loyal; some who moved away continued to return from as far away as California for their checkups. The whole town felt involved, and so an aroused community, including religious and civic groups, PTAs, business, industry and private individuals, rallied to raise the money to keep the program going. Boston University offered to take it over and administer it under the same leadership, retaining the services of its first director, Dr. Thomas R. Dawber, and his long-time colleague, Dr. Kannel, serving as Director. New funds came from Washington, and a new phase of the study began.

The rate of new coronary attacks after 1968 was 176 per 1,000 among persons with an HDL under 25 mgs. By contrast, it was only 60 per 1,000 among those with an HDL of 45 to 54 mgs. The "new" Framingham is also the first where some of the offspring of the original participants are under scrutiny at about the same age their parents were when they entered the study — a rare opportunity to observe the influence of both genetics and environment on coronary heart disease. "We examined more than 2,000 offspring (ages 12 to 62) between 1971 and 1975," Dr. Manning G. Feinleib reported at the American Heart Association meeting in the fall of 1976. "Among them were 374 whose fathers had developed coronary heart disease." How did the offspring show up as risks? "The one consistently high risk factor they exhibited was a *low* HDL, varying with the age of the father at the time of his first attack. When the father was a victim under the age of 50, the son's HDL was significantly lower than those of the offspring whose fathers' attacks were between 50 and 65. One of the ironies is that among these high-risk sons of high-risk fathers, there was a *larger percentage of smokers* than among sons of fathers with no coronary heart disease."[28]

Reports from San Francisco, Albany, New York, Evans County (Georgia), and Honolulu tell the same HDL story that is coming out of Framingham — the higher your HDL, the smaller the risk of premature coronary heart disease.

Today, the most dramatic development in promoting longevity is the concept that sudden death need not be lethal. When the Roman scholar, Pliny, wrote about "sudden death" almost 2,000 years ago (ironically, shortly before his own sudden death while attempting to escape from Pompeii during the eruption of Vesuvius in A.D. 79), he chronicled cases of those who died suddenly from sorrow, shame and joy, as well as some who were doing nothing more unusual than putting their shoes on in the morning. The circumstances under which sudden death strikes today are not too different. In the last ten years, however, for the first time in history, it has become possible to (1) identify many potential candidates in advance; (2) decrease the odds of sudden death through treatment with an increasing choice of effective drugs; (3) bring the arrested heart back to life with immediate and on-the-spot help, often by a passerby, a colleague or a relative.

[28] In a further effort to separate genetics from environment, Dr. Feinleib is now studying adopted children, children of previous marriages, and 1,800 spouses of their offspring.

"In the industrially developed world, sudden death looms as the leading cause of fatality," says Harvard University's Dr. Bernard Lown. "Almost two-thirds of coronary heart disease deaths occur within twenty-four hours after onset, and in the majority of those dying suddenly, death is unexpected, occurs instantaneously, is unheralded by symptoms, and strikes outside the hospital while the victim is engaging in usual activity."

The fact is that many who die suddenly do not have evidence of a coronary thrombosis (blood clot in a coronary artery) or a recent myocardial infarction (destruction of heart tissue).[29] Their collapse is caused by dangerous disturbances in the electrical impulses that normally regulate the heartbeat. The outcome of the "electrical storm" is ventricular fibrillation, a condition in which the large muscular chamber of the heart contracts in a chaotic and uncoordinated manner. The heart stops pumping, and cardiac arrest may ensue. "The electrical accident," asserts Dr. Lown, "is both reversible and preventable."

The concept that cardiac arrest need not mean death and that the heartbeat can be restored with prompt treatment developed in the 1960s. It was a lesson learned from the newly installed coronary care units, when monitors rang the alarm for the patient whose recent infarction may have brought on the arrhythmia. The odds for resuscitation in such a setting are understandably high, but since more than half of all heart attack victims died before

---

[29] When Elvis Presley, the first King of Rock, died suddenly at the age of 42, in the summer of 1977, his death was attributed to a cardiac arrest, brought on by arrhythmia. But contributing to his vulnerability were several years of irregular heartbeat, high blood pressure, diabetes, overweight, stress (the crown was not always secure on his head), and a family history of premature CHD. Nineteen years earlier, almost to the day, his mother had died of a heart attack — also at 42.

reaching the hospital, care had to be provided at the scene. The first on-the-spot program was organized by Dr. J. F. Pantridge in Belfast, Northern Ireland. His flying squad of specially trained and equipped medical personnel were transported to the victim to supply the emergency care needed.

The pioneer in the United States was Seattle, Washington, where today more than 100,000 of the 500,000 population have completed training in CPR (cardio-pulmonary resuscitation), the life-saving technique that combines mouth-to-mouth resuscitation and cardiac compression. Paramedics and fire department personnel are trained and respond to a call within five minutes of receiving it. But the real heroes in resuscitation are the passersby who apply CPR within less than five minutes of the arrest itself. They have saved lives in supermarkets, restaurants, conventions, city streets and city halls. Among the heroes are policemen, teachers, clerks and housewives. One beneficiary owes his life to an insurance executive who had just completed an American Heart Association CPR course.

The advantage of "early" (less than a five-minute lapse) intervention over "later" trained rescue squads was stressed in a report to the American Heart Association in November 1976. Said University of Alabama's Dr. Donald P. Copley: "There was a dramatic difference in their conditions when they arrived in the hospital. Not only did more survive, but they were more alert, and, fortunately, none suffered mental impairment."

CPR has caught on all over the country. In fact, when Sally Miller tried to sign up for a Red Cross–sponsored course at her local high school, she found the first class fully subscribed and was wait-listed for the succeeding one.

Key to prevention of sudden death from heart attack is

the early identification of the person at risk, and the growing knowledge of how to deal with the risks. Most susceptible are survivors of one or more myocardial infarctions. The greater the damage to the heart muscle, the greater the risk. Limiting the size of the infarction soon after the attack is now the target of intensive research, which is yielding very promising results today. Another clue to electrical instability is a particular type of ventricular premature beat, PVB, which shows up on the electrocardiogram. Drugs that act on the deranged conduction pathway are now controlling dangerous arrhythmias in many.[30]

"To catapult the electrically unstable heart into ventricular fibrillation requires other inputs," says Dr. Lown, and, in his opinion, the other inputs come from the nervous system and neurohormones (catecholamines). Among the factors at work in releasing neurohormones are two familiar risk factors for coronary heart disease — smoking and stress.

As far back as 1968, Drs. Kannell and Castelli reported that sudden death occurred five times more often in heavy smokers than in non-smokers. Indeed, sudden death in non-smokers is a rare event! Several constituents of cigarette smoke have been implicated, notably carbon monoxide, which cuts down on the amount of oxygen available to the heart, and nicotine, which increases the secretion of catecholamines. With the recent development of highly sensitive methods for measuring catecholamines, it is now possible to document the consequences of smoking. Within *ten minutes* of lighting up, plasma catecholamines rise, pulse rate quickens, blood pressure

---

[30] Still to be perfected is a pacemaker that will control the erratic beat as proficiently as the familiar pacemaker now speeds up slow heart rates.

increases, and the normal ratio of the blood constituents is altered.[31]

Although it is not yet possible to nail down the contribution of stress, emotion and behavior to coronary heart disease with the same hard data as a cholesterol level or a blood pressure reading, a picture is beginning to emerge that goes far beyond the anecdotal experiences reported down through the ages. More than two hundred studies in the last twenty years have implicated such hard-to-measure factors as personality, change of residence, change of job (or more crucial — loss of job), and stressful life events, but it is only in the last five years that these factors can be measured.[32]

The coronary-prone personality and his behavior were succinctly described by Dr. Osler back in 1910 in an address to the Royal College of Physicians: "the robust, vigorous in mind and body, keen and ambitious man . . . whose engine is always at full speed ahead." Today, thanks to studies initiated on the West Coast in 1959 by San Francisco cardiologists Drs. Meyer Friedman and Ray H. Rosenman, the case for the coronary-prone personality — the Type A (in contrast to the less vulnerable Type B) — is gaining increasing acceptance even among skeptics in medical circles. He is often deeply involved in his job or profession, competitive, aggressive, impatient and always fighting the clock. He is in rigid control of his emotions, and while he may have a great capacity for enjoyment, he is apt to lack spontaneity. There is a powerful element of denial in his life. One study of 345 survivors

---

[31] Smoking also contributes to coronary heart disease by increasing the stickiness of blood platelets — an important factor in the formation of a blood clot.

[32] Covered in depth by Boston University's Dr. C. David Jenkins in a two-part series in the *New England Journal of Medicine*, May 1976.

of myocardial infarction showed that 20 percent were denying that they even had had a heart attack within three weeks of the event.

The validity of the profile was strengthened when it was applied by Drs. Friedman and Rosenman to persons who were at risk but had not yet been affected by coronary heart disease. In the 1960 Western Collaborative Group Study, more than 3,500 men, ages 39 to 59, who were free of coronary heart disease were rated A or B solely on the basis of their behavioral traits. After eight and a half years of follow-up surveillance, the Type A men exhibited an overall twofold risk, after taking into account cholesterol, blood pressure, cigarettes, weight and age. Among younger men between 39 and 49, the incidence of coronary heart disease was six times higher among the Type A than among the more relaxed, easy-going Type B. Their attacks were more severe, more frequent and more likely to be fatal.

Moreover, coronary-prone behavior often has its origin early in life, it was learned from a study of 120 teenage sons of Type A fathers. At 15, they showed a striking similarity to their fathers — a pattern more likely imposed by the father's expectations and behavior than by his genes.

Although women were not included in the prospective (predictive) studies, retrospective studies reveal that among them the coronary-prone behavior pattern is also associated with an increased prevalence of coronary heart disease. Not unexpectedly, a recent study[33] suggests that this behavior pattern may be more common among employed women than among housewives; more common

[33] Ingrid Waldron, Ph.D., *et al.*, "The Coronary-Prone Behavior Pattern in Employed Men and Women," *Journal of Human Stress*, December 1977.

among men than among women; and more common among men high on the Type A scale (aggressive and competitive men) than among those lower on the scale.

Later studies added to the original picture: the coronary-prone is subject to anxiety, depression (not easily discerned), sleep disturbances, fatigue and emotional drain. Similar observations (based on individuals who have suffered heart attacks) have been made by investigators in Australia, the Netherlands, Israel and Sweden.

"Coronary-prone behavior pattern is not the same as stress," explains Dr. Jenkins. "It is a style of behavior with which some persons habitually respond to circumstances which surround them." And stress is not equally hazardous to all, for many meet the challenges well and even thrive on them. But when a life event is beyond the individual's power to control and his ability to cope, the cardiovascular system becomes a vulnerable target.[34]

"Broken Heart: A Statistical Study of Increased Mortality Among Widowers" is an account in the *British Medical Journal* of 4,486 widowers who were 55 years old or older when their wives died. Dr. C. M. Parkes and his colleagues followed them for nine years. The death rate among them during the first six months of their bereavement was 40 percent higher than expected for married men of the same age, with the greatest contribution to mortality — coronary heart disease — 67 percent higher than expected.

Among the vulnerable, changes in lifestyle and mode of earning a living play a powerful role in development of

[34] The quantitative contribution of stressful life events has been receiving an increasing amount of attention over the past ten years. A Social Readjustment Rating Scale, developed by Drs. T. H. Holmes and R. H. Rahe, to measure the impact of the death of a loved one, change of residence or job, birth of a child, injury, etc., has been refined and used extensively, but there are still limitations to its predictive value.

coronary heart disease. Early studies in North Dakota and California revealed that the rate among men who had experienced several lifetime job changes and geographic moves was twice as high as that of men who had experienced no such changes. The rates were three times as high among men reared on farms who later moved to the cities and took white-collar jobs. In each of these studies, the differences between the victims and their healthy counterparts could not be attributed to the usual risk factors — diet, smoking, blood pressure or family history. More recent studies in the South and other parts of the country tell the same story. Upward mobility takes its toll, as does a change in cultural setting. A 1973 report in the *Journal of Chronic Diseases* discusses a higher incidence of myocardial infarction among persons whose fathers immigrated to Israel than among those whose fathers were born there.

An unusual opportunity to study "occupational stress" developed at the Kennedy Space Center, where every successful space launching — and the stress accompanying the accomplishment — was followed by a 15 percent cutback in budget and personnel. University of Nebraska's Dr. Robert S. Elliot and Harvard University's Dr. Herbert Benson described what happened to personnel over a period of eight years when their jobs were threatened. Incidence of sudden death or acute myocardial infarction among 200 men selected at random, with an average age of 31, was 50 percent higher than that of a similar group of controls taken from the general population of Florida. Moreover, the usual risk factors — elevated blood cholesterol, diabetes, obesity, cigarette smoking — were not present. "The only consistent finding was occupational stress," the researchers told the 1977 meeting of the American College of Cardiology.

Discussing the link between the brain, stress, and heart attack, Drs. Elliot and Benson explained that the surge of catecholamines through which your body responds when you are under great stress pushes the heart to perform beyond its capability, resulting in a type of cell death of the heart muscle in which it is literally "worked to death." This is quite different in nature from the heart damage heralded by chest pain or other familiar warning of a myocardial infarction.

For the most part, however, the physiology of stress has a direct link to the physiology of coronary heart disease — the hormones of the nervous system affect fat metabolism, salt regulation and blood clotting, as well as the metabolism of the heart muscle itself. Racing-car drivers in the Grand Prix, medical students preparing for an anatomy exam, even healthy young men behind the wheel of a car in heavy traffic experience elevated serum cholesterol and other fats and accelerated clotting time.

Still another clue to the role of emotion and stress in coronary heart disease is the success of a group of drugs (beta adrenergic blocking agents) that interrupt the regular pathway between the brain, the heart and the blood vessels. These drugs are now being used in the treatment of high blood pressure and angina.[35] Dr. Lown and his Harvard colleagues succeeded in protecting animals with myocardial infarction from being further damaged under stress by pre-treating them with the blocking drugs before exposing them to a stressful situation that, in the past, would have thrown them into ventricular fibrillation.

Dr. Lown's success is not limited to the laboratory. Among his patients is a 40-year-old man who has no evidence of heart disease but has been brought back from the brink of death twice, the first time by his wife, who is

---

[35] A severe chest pain brought on by exertion and caused by an insufficient amount of oxygen in the heart. Unlike the pain of a heart attack, the pain of angina subsides on resting and rarely lasts more than half an hour.

trained in cardiac pulmonary resuscitation. Both cardiac arrests were associated with highly emotionally charged and stressful situations. To control the fibrillation and premature beats that could trigger a third arrest, Dr. Lown is using a combination of drugs, plus a remedy that is commanding a growing respect in the cardiovascular field — meditation.[36]

The search for a healthy heart leads aware Americans down many paths, some more rewarding than others. Not too long ago, a story was circulating that the eggs of the Aruacana fowl (the Easter-egg chicken) are low in cholesterol. Not so, reported scientists who conducted independent studies at six United States universities, "providing evidence to disprove the myth."

Recent enthusiasm for the value of a high-fiber diet, particularly in lowering blood fats, was dampened by researchers at the University of Oregon, reported Dr. Thomas L. Raymond to the 1976 Scientific Sessions of the American Heart Association. In a highly supervised, well-controlled study, one group was fed a cholesterol-free diet, the other high cholesterol, including four egg yolks a day. Predictably, the cholesterol levels of the first group fell and the egg-fed subjects' cholesterol rose. When high-fiber foods derived from corn, beans, wheat and citrus fruits were added to the menus of each group, the low cholesterols did not become lower and the high cholesterols remained high.

"We do think that there are beneficial effects from high fiber, but not necessarily toward heart disease," Dr. Raymond concluded. "The four eggs a day are the bad part; you can eat all the fiber you want and it's not going to help you — it won't dissolve cholesterol."

[36] See Chapter 3.

Two popular vitamins — C and E — are also making news in relation to coronary heart disease. Claims that vitamin E can relieve the pain of angina were not fulfilled in a study reported to the 1977 Public Health Service Professional Association meeting. Even large doses over a fourteen-month trial "failed to increase exercise capacity, improve left ventricular function, or reduce frequency of pain," Dr. Dennis G. Caralis, cardiologist, Baltimore United States Public Health Service Hospital, said. On the other hand, the relatively large doses — 1,600 units of vitamin E daily — were not harmful, an important bit of news, given the easy availability and widespread use of the substance.

The case for vitamin C is also controversial. Experimental pathologist Dr. S. D. Turley and his colleagues at the Australian National University urge that the time is right for wide-scale studies.[37] Their enthusiasm is based on data suggesting that a widespread subtle vitamin-C deficiency is contributing to high blood cholesterol, which can be reversed, they claim, by larger but not massive doses of vitamin C. On the other hand, a recent review of the field by NHLBI scientists finds no demonstrable benefit.

Probably no recommendation in a regimen of protection against heart disease will give you greater and more instant gratification than exercise. You will feel better, look better, relax more and almost certainly achieve added success in bringing and keeping your weight down. The therapeutic aspects, however, are not so easy to document.

The most passionate advocates are found among the 10,000 Americans (including 1,000 physician members of the American Medical Joggers Association) who are veterans of the grueling 26-mile, 385-yard marathon race.

[37] *Atherosclerosis*, July-August 1976.

Their enthusiasm received a big boost with the recent disclosure of the coveted increased HDL.[38] "Slow long-distance running, combined with no smoking and a diet low in animal fats, salt, refined food, liquor and wine, confers absolute protection against fatal coronary heart disease," asserts California pathologist Dr. Thomas J. Bassler. The widely publicized promise of immunity to the ravages of atherosclerosis is attracting an increasing number of new recruits to long-distance running. Many are older men and women, some of whom have recovered from heart attacks. But despite Dr. Bassler's unqualified faith, his contentions are by no means unanimously accepted.

The safety and efficacy of marathon running was debated in depth for the first time in the fall of 1976, when the New York Academy of Sciences brought together doctors and scientists (about one-third marathoners) to look at its medical, psychological, physiological and epidemiological aspects. The time and place were no accident, for Dr. Paul Milvy, Mt. Sinai School of Medicine biophysicist and a 3-hour, 22-minute man, began planning the meeting a year in advance to coincide with the first annual Five Borough New York City Marathon, which attracted more than 2,000 entrants, 459 of whom were over 40, the oldest being 71. (None suffered a heart attack during the race.)

There was no consensus at the end of the four days of discussion, but tentative guidelines began to emerge, placing the divergent claims in perspective. Some of the comments by participants:

☐ Dr. Tim Noakes from South Africa, a 3-hour, 8-minute New York City marathoner, discussed five cases of heart attacks in men who had completed marathons, including

[38] See page 47.

that of a 35-year-old, a ten-year veteran of marathons, who died. All developed symptoms while running and continued to run.

☐ Dr. Terrence Kavanaugh (a 4-hour, 15-minute man) of Toronto, who directs a program of long-distance running for post-coronary patients, "I would be very hesitant to say, carte blanche, that all patients who have had a coronary should . . . try to run a marathon. . . . Furthermore, I am convinced that the more people over 45 who run the marathon, the more deaths Tom Bassler will begin to pick up." Dr. Kavanaugh also deplores the new competitive aspects of running: ". . . it's going to kill a lot of the tremendous benefits that could come from a good, active, healthy life."

☐ New Jersey cardiologist Dr. George A. Sheehan: "This meeting says to me, 'Just because you can run ten or fifteen miles does not mean you can't still have a coronary.' "

☐ Dr. Milvey comments that "the overwhelming majority of the 10,000 marathoners are thin, do not smoke, drink sparingly, with a minimum amount of meat in their diets. If a similar group of 10,000 who *don't* run marathons were studied, their expected coronary rate might not be too different from the runners'."

But would they have as much fun? Harvard University's Dr. Herbert Benson remarked in a recent interview that long-distance runners describe a relaxed, almost euphoric state, not too different from the meditative state, after about ten miles. And our 27-year-old son, a veteran of five marathons, confirms the euphoria.

Despite the glamour of the marathon, only a small minority of men and women will attempt such a long distance. If you choose jogging for your favorite sport, ten miles a week would be more realistic. Even so, says a widely recognized expert in exercise physiology, Dr. Jere

Mitchell, Professor of Internal Medicine at the University of Texas Southwestern Medical School, "there is no conclusive scientific evidence that exercise prevents heart disease . . . or increases longevity . . ."

The scientific evidence is still contradictory. A recent study by a Finnish researcher who compared men in executive-type sedentary careers with men who had been cross-country skiers all their lives found no difference in longevity. On the other hand, a twenty-two-year follow-up on 3,686 San Francisco longshoremen by University of California's Dr. Ralph S. Paffenbarger, Jr.,[39] disclosed that heavy workers — such as cargo handlers, whose duties require repeated bursts of high energy output — had a coronary death rate almost half that of medium, light and sedentary workers. Sudden death was 75 percent less among the heavy workers.

"You can make a good case that exercise improves the quality of life without having to make any false claims that it increases the quantity of life," says Dr. Mitchell. "There is evidence, however, that exercise produces a general feeling of well-being and can reduce depression and hypochondria, especially in patients who have had a heart attack." Indeed, for David Miller, the confidence he acquired from being able to climb two flights of stairs, ride a bicycle to town for the Sunday paper, and walk to the station in the morning was enough to dispel his fear of resuming a sex life.

Many doctors, reluctant to be unnecessarily restrictive when asked by a post-coronary patient "when?" and at the same time uneasy about being totally permissive, will answer evasively. For them, a February 1976 editorial in the *Journal of the American Medical Association* brings the

---

[39] Who completed a seventy-two-mile marathon around Lake Tahoe in 11 hours.

known data up to date. The better a patient is trained to tolerate a work load, the slower is his heart rate during coitus. The energy cost (and strain on the heart) during coitus are no greater than performing usual occupational activities. Yes, some men *do* die during coitus, but when that happens, it is apt to follow a pattern of a "clandestine rendezvous, in unfamiliar surroundings, after a big meal with a few drinks." The inference could be that guilt contributes.

Some do's and don'ts on exercise from Dr. Mitchell: Start out on a program slowly, and if you are over 35, see a physician and have an exercise stress test. If your goal is to build up skeletal muscles (arms, legs, etc.), isometrics (static exercises such as weight lifting, where you control muscles without moving limbs) is fine. If, however, you are looking to improve cardiovascular fitness, isometrics will not only be useless, they may be dangerous. They raise blood pressure and put a strain on the heart. Persons who die changing a tire, shoveling snow or straining to open a stuck window are involved in isometrics. Choose instead isotonic dynamic activities (moving your limbs) — swimming, walking, jogging.

The landmark advances in prevention of atherosclerosis in the 1960s have been paralleled by progress in reversal of atherosclerosis in the 1970s. In the summer of 1977, Dr. David Blankenhorn and his colleagues at the University of Southern California reported regression of plaques in the femoral (thigh) arteries of heart patients 40 to 49 years old, all Type A, hard-driving, heavy-smoking business executives. All gave up smoking completely after the attack, and in a little over a year, the comparison of before and after measurements of the plaques revealed significant regression. Earlier in the year, the California team stirred the medical world with an announcement of their success in demonstrating, for the first time, regression of human atherosclerosis in post-coronary patients brought about by lowering abnormal blood fats (cholesterol and triglycerides) with diet, drugs and exercise.

With the cooperation of NASA scientists at the Jet Propulsion Laboratory, California Institute of Technology, the team is now utilizing the advanced computer technology developed for analysis of photographic images taken by spacecraft. Because the technique is sensitive to small changes, it enables scientists to visualize the plaques and measure hitherto unmeasurable changes over a period of time. This study[40] is also a first to dispel the doubts of many investigators that reversing risk factors will indeed bring about reversal of established atherosclerosis.

Only patients whose abnormal cholesterol, triglycerides and blood pressure were lowered showed regression. Where there was no fall in these risk factors, there was no improvement in atherosclerosis. Emphasizing that the number of patients is still small, Dr. Blankenhorn finds the greatest hope for those who are relatively young, and in whom the atherosclerosis is not too advanced.

At the University of Minnesota School of Medicine, Dr. Henry Buchwald has devised an operation — the partial ileal bypass — designed to interfere with absorption of cholesterol in the intestinal tract and speed up its elimination from the body. While early results seemed promising, later attempts both by Dr. Buchwald and other surgeons have proved disappointing.

The most convincing evidence that atherosclerosis can be successfully treated continues to come from animal studies. Rhesus monkeys with advanced lesions experienced remarkable improvement in all of their coronary arteries after forty months of a low-fat, low-cholesterol

[40] *Annals of Internal Medicine*, February 1977.

ration or high-corn-oil, low-cholesterol diet.[41] Atherosclerotic swine fed a low-fat mash diet for fourteen months, during which time their serum cholesterol declined to normal, experienced a striking improvement in their coronary arteries.[42] Further evidence of successful reversal comes from angiograms on dogs — before, during and after therapy.[43]

The saga of the identification of the culprits in atherosclerosis is not unlike a suspense story where a strong case is built up against a major suspect, and before the dénouement, a relatively minor figure or a new character is suddenly catapulted into the spotlight. The scenario dates back more than a century to the German pathologist Rudolf Virchow, who proposed that atherosclerosis starts with injury to the inner lining of the blood vessel (endothelium) associated with the deposit of cholesterol, an irritant that produced "inflammation" in the blood-vessel wall. The 1856 hypothesis became increasingly attractive as twentieth-century observations on the worldwide connection between cholesterol, fat, atherosclerosis and coronary heart disease began to pile up. Indeed, the atheromatous plaque, a discrete fibrous lump, is so rich in fatty substances that cholesterol crystals can often be seen without the aid of a microscope.

Concepts began to change, however, when the superior magnification of the electron microscope revealed that the plaque is not the usual response to an injury — but that it contains a large number of smooth muscle cells normally found only in the blood-vessel wall. How did they get out and why? In the light of this new puzzle, the original

[41] M. L. Armstrong *et al.*, *Circulation Research*, 1970.
[42] A. S. Daoud *et al.*, *Circulation*, 1974.
[43] R. G. de Palma *et al.*, 1975.

scenario is now being rewritten, presenting an entirely new approach to the origin and treatment of atherosclerosis.

The first step is in the endothelium, a single layer of cells lining the inside of the blood vessel, covering it like a thin carpet. Intact, it acts as a barrier, keeping most substances in the circulating blood (including lipoproteins) out of the vessel while it speeds the blood on its way. Facilitating the smooth journey is not only the super-teflon nature of the lining, but also a recently discovered substance synthesized by the endothelium — prostaglandin PGX — that interferes with clot formation.

When the barrier is breached, the layer under it is destroyed, and platelets — the small irregular disc-shaped bodies in the blood necessary for formation of blood clots — invade. No longer inhibited by the endothelium, the platelets can now clump together and adhere to the exposed tissue. They inflict further damage by releasing a substance that stimulates the normally quiet, smooth muscle cell to multiply rapidly. Now lipoproteins also invade, and conditions are ripe for the beginning of an atheromatous plaque — with the newly grown smooth muscle cell a required ingredient, plus cholesterol and other fats, cell debris, connective tissue, etc.

The initial injury may be caused by the familiar risk factors — too much fat in the blood, high blood pressure pounding away, toxic substances from cigarette smoke, infection, neurohormones, etc. — but the lesion is not permanent unless the assault is repeated and prolonged. Caught in time, the lesion can be reversed; in 1976, not too long after the platelet connection was established, an anti-platelet serum was developed. The new look of atherosclerosis comes from research on baboons, rabbits, and pigs, but at least one investigator promises that "what happens in animals will relate to humans."

In the late summer of 1977, we had the opportunity to spend an evening with scientists from three laboratories where this trail-blazing work is now going on.

☐ Dr. Lawrence A. Harker, whose colleagues at the University of Washington School of Medicine, Drs. Russell Ross and John A. Glomset, focused attention on the endothelium in 1973, is a member of the team that developed the next step — the platelet role. Their subjects are baboons and pigtail monkeys. From them, they learned about the potential to reverse atherosclerosis. "After a single episode of injury to the epithelium, the lesion grows for three months. By six months, it regresses . . ."
On aging and coronary heart disease: "A possible clue to the greater risk of CHD with aging is our observation that old primates have many endothelial cells missing. Not so in younger animals." The next step: "Our immediate task at all ages is to enhance the integrity of the endothelium with medication, and there is hope that such a drug can be found."

☐ Dr. Valentine Fuster, working with Dr. E. J. W. Bowie, Mayo Clinic, turned to pigs with a platelet defect to learn the effect on atherosclerosis. The condition is von Willebrand's disease, a bleeding disorder related to hemophilia. Two groups of pigs were fed diets equally high in cholesterol. Those free of the bleeding disorder (with normal platelets) developed severe raised plaques of atherosclerosis. The von Willebrand pigs got off with flat (mild) atheromata. "We are starting now to look at von Willebrand victims. Do they, too, escape coronary heart disease? We should have an answer in five years."

☐ Dr. Theodore A. Spaet, working with Dr. R. J. Friedman at New York City's Montefiore Hospital, where the benefit of anti-platelet serum is undergoing intensive scrutiny, talked about its success with rabbits.

"Given to the animal three days *before* the endothelium was injured, it provided good protection against atherosclerosis. When the anti-platelet serum was administered one day *after* the injury, it was too late. The muscle cells had already received the message to grow out. The formation of the plaque was set in motion." Dr. Spaet ended on a note of confidence: "Success in reversing atherosclerosis will escalate in the next few years."

Still another hypothesis on the origin of atherosclerosis comes from the University of Washington's Drs. Earl P. Benditt and J. M. Benditt, who propose that each plaque starts as a benign tumor of a single smooth muscle cell that has proliferated because it has undergone a mutation. Since confirmed by other investigators, this theory has been recently hailed as "one of the most significant observations made in the entire field in the last twenty years." [44]

"The idea that the atherosclerotic plaque may be some form of neoplasm . . . is quite startling if one's concept of a neoplasm is limited to malignant cancers that spread," Dr. Benditt explains. "Many tumors, however, are benign; they remain localized, grow slowly, and may even regress." His evidence that it starts with a single cell comes from an ingenious application of chromosome and enzyme studies used in a variety of genetic disorders.[45]

One of the most attractive features of the monoclonal (single cell) hypothesis is that it provides a new framework within which one can ask new questions about the role of various risk factors, and Dr. Benditt explains how it may work with one of the most powerful and least understood risks. Cigarette smoke contains precursors of mutagens such as aryl hydrocarbons. Because they are fat soluble,

[44] Dr. Robert H. Heptinstall, Johns Hopkins University School of Medicine.
[45] Discussed in detail in *Scientific American*, 1977.

they are carried in the blood in the lipoproteins — the higher the lipoprotein, the more aryl hydrocarbon can be put in contact with the smooth muscle cell of the heavy smoker. And the other two big factors also fit into the monoclonal theory: high-cholesterol and high-fat diets have been associated with mutagenesis (cancer) in the intestinal tract,[46] and the DNA of people with hypertension is more susceptible to breakage in cells by mutagens than is the DNA of people with normal blood pressure.[47]

Acceptance of Dr. Benditt's theory is not universal, and he acknowledges that it "does not simplify the problem of identifying the causes of heart attacks and strokes . . . but it puts us in a much better position from which to consider, test and identify the multiple factors, both genetic and environmental."

Putting together data on preventing first heart attacks and cutting down the damage that could lead to a second is now recognized as a major public health problem in the countries most profoundly affected. Unfortunately, protecting the public, particularly from a disease where your behavior today will ward off trouble in ten, fifteen or twenty years, has, in the past, been a singularly unrewarding exercise. There was small enthusiasm within the profession and large apathy among the public. No longer!

Finland, with the worst heart attack rate of all developed nations, mounted a five-year prevention program in 1972, directed at all 180,000 residents of the rural eastern county of North Karelia, where the proportion of people who died from heart disease was larger than anywhere else in the world. Although 70 percent of the population do farm work or forestry, their physical activity

[46] Dr. Ernst Wynder.
[47] R. W. Pero, University of Lund, Sweden.

failed to protect them from the effects of excessive smoking, elevated blood cholesterol, and high blood pressure. A variety of projects, financed by the government with assistance from the World Health Organization, involved the entire community. Changing the diet was the most difficult task. In place of the thick butter they loved to smear on bread, they reluctantly turned to margarine. The dairy industry promoted low-fat milk, and eventually non-fat milk. Grilled sausages (a popular delicacy after a sauna) that were formerly fatty are now made with 25 percent mushrooms. Fresh vegetables, never popular in the past, are now being grown in backyards.

Laws were passed forbidding smoking in public buildings and transport, and widespread publicity in the media, leaflets and posters saturated the county. Success in lowering blood pressure has been far less dramatic, but the percentage of hypertensives receiving treatment has increased appreciably, indicating that the disease is being caught earlier, which facilitates better treatment. By 1977, the heart attack rate had declined 40 percent for males, the thirty-year annual increase had stopped, and North Karelia had dropped to fifth place among Finnish counties in incidence of coronary heart disease.

In Britain, a 1976 report of the Joint Working Party of the Royal College of Physicians of London and the British Cardiac Society on Prevention of CHD formulated a body of advice for the medical profession of the country. It includes the now-familiar reminders on diet, smoking, blood pressure, diabetes and a few more. On stress, the report says: "Initiative, diligence, hard work, especially in young people, should not be discouraged on the mistaken supposition that these qualities are indicators of future coronary heart disease." On oral contraceptives: "Use with caution in women over 40, those with a family history of premature CHD, those who smoke more than

twenty cigarettes a day, or have other risk factors." On
children: "Measures . . . apply as much to children as to
adults since all the major risk factors found in adult life can
occur during childhood." On responsibility: ". . . implications for a national food policy . . . for producers and
manufacturers of food . . . regulations for labelling . . .
review practices in schools, hospitals, armed forces, etc."

In December 1975, the Ministry of Agriculture of
Norway presented a government white paper outlining a
proposal for a comprehensive food and nutrition policy.
Since hailed as a milestone, it spells out specific nutritional
goals: less fat, meat, whole milk, eggs and sugar, and
more chicken, fish and grains. Furthermore, to encourage
the production and consumption of desirable foodstuffs,
production subsidies will be used as much as possible to
aid desirable domestic products, and consumer subsidies
to bring down prices of desirable foods fixed in the international market.

It is ironic that in the United States, where much of the
data on risks and coronary heart disease was developed,
"the policymaking arms of the Federal government have
come up with no national nutrition policy containing specific nutritionally oriented goals for the American public."[48] To be sure, dietary goals have been formulated and
consistently advocated by the Senate Select Committee on
Nutrition and Human Needs, but it has never had the
power to implement its goals. The first legislative follow-
up came in February 1977, when Senator Hubert Hum-
phrey introduced a bill addressing some of the nutritional
issues underscored by the Committee's publications (May
1975, January 1977).

Do we need a cradle-to-grave dietary program? On the

[48] Dr. Beverly Winikoff, Health Sciences Division, The Rockefeller
Foundation, *American Journal of Public Health*, June 1977.

dubious side is Rockefeller University's Dr. E. H. Ahrens, Jr.: "Any advice to the general public to make large dietary changes now is considered premature."[49] On the other hand, Dr. Jeremiah Stamler of Northwestern University remains a staunch believer that the time is long overdue to implement a change for adolescents as well as adults. "If we changed the dietary patterns of the population, we might be able to eliminate the coronary heart disease problem," he predicts. Indeed, Dr. Levy's July 13, 1977 press conference, at which he announced the news of the waning CHD epidemic, lends support to Dr. Stamler's contentions.

Meanwhile, the National Heart, Lung and Blood Institute is grappling with the problem of preventing CHD through programs involving persons at very high risk. More than 500,000 men between 35 and 59 were given free cholesterol tests at work, football games, shopping centers, Red Cross blood banks, etc. From among them, 3,800 men with the following characteristics were chosen in 1976: (1) high LDL and more than 265 milligrams cholesterol; (2) absence of heart disease; and (3) a willingness to remain in the program for seven years. One group will be treated with diet and cholestyramine, a cholesterol-lowering agent; the other with diet and a placebo. Follow-up examinations and evaluations will be handled by twelve Lipid Research Clinics across the United States and in Canada.

Also in 1976, a more challenging prevention program got off the ground. From a group of 366,000 men between 35 and 57 who were screened, 12,500 who possessed one or more of the big three risk factors — elevated cholesterol, high blood pressure, cigarette smoking — were selected to participate in a clinical trial designed to reduce

[49] *Annals of Internal Medicine*, July 1976.

or eliminate these risks with intervention and education. The Multiple Risk Factor Intervention Trial — MR FIT — will run for six years, and the recruits promise to remain for the duration. "If the intervention proves effective in men," says NHLBI director Dr. Levy, "it would also be applicable to prevention of CHD in women, since the same factors affect CHD risk in both sexes."

During a visit to the American Health Foundation, where the first MR FIT began to function, psychologist Dee Burton, Director of Intervention, described how the program is progressing. Because they started two years before the last MR FIT center made its final selection, they will have eight years to lower the risks in their participants. Who are they? All are from the New York metropolitan area; they came from the Sanitation Department, the City Health Department, City University, and in response to radio, TV, and *New York* magazine ads.

The test group participates in ten weekly group meetings to which they are encouraged to bring their wives. Meeting with them at various times are a psychologist, a nurse, a hypertension specialist, nutritionists, a health counselor, and an individual case manager for each. "The meetings are not all passive," Ms. Burton explains. "They all share their experiences, and the college professor often learns from the policeman. As far as attitude and behavior change [which all promise when taken into the program], this interchange is very helpful." Further, she adds, "they can deal better with handling their risk factors after they have had an opportunity to deal better with other problems in their personal lives."

While it is still too early to evaluate the effect of the program, initial signs are good. Compliance in control of blood pressure is excellent — participants are sustaining loss of weight and cutting down on salt and alcohol.

Success in cutting down smoking (in a program which includes relaxation with the help of a recorded tape) is 25 percent, which is high compared to other programs. Men who live alone have learned to cook — one even graduated to gourmet cooking. A recurrent theme is that they must take responsibility for their own health. "And to reinforce the education and training, we have a good social program for the participants and their wives."

A total of 4,500 men and women, ages 30 to 69, all of whom have suffered one or more documented heart attacks in the last five years (and are free of other major diseases), are participating in an NHLBI study designed to determine if the regular use of aspirin will reduce the risk of another heart attack or a stroke. The rationale is that aspirin is known to interfere with formation of a blood clot by coating platelets,[50] which reduces the stickiness essential for clot formation.

Documentation for such a hope comes from the Boston Collaborative Drug Surveillance Program, Boston University Medical Center, whose scientists have been conducting intensive monitoring of a variety of drugs since 1966.[51] In that study, fewer than 1 percent of all patients with myocardial infarctions were regular aspirin users, in contrast to almost 5 percent for the non-users. The program will not be complete until the end of the decade, but early unofficial results are said to be disappointing, because aspirin, it has since been discovered, works on other aspects of platelet function as well. It may turn out that for aspirin to produce its beneficial effects, a critical dose is required.

There is no doubt, however, about the benefit to the

[50] See page 69.
[51] In various hospitals in the United States, Israel, Scotland, Italy and Auckland.

heart attack victim (or one suspected of CHD) of the newly developed ability to look into the heart and blood vessels, both at work and at rest, to differentiate between old and new damage, and, above all, with increased safety for the patient. Until now, reliance has been on angiography — where an opaque dye is injected into the coronary arteries and an x-ray motion picture is taken.

While it takes a good picture of the blood vessels, what it tells about damage to the heart muscle itself is limited. Moreover, because it is an invasive technique, it is not without its hazards.

Echocardiography directs an ultrasonic beam to the heart with no discomfort to the patient, is useful in detecting early damage before symptoms appear. More sophisticated and more revealing is another non-invasive technique,[52] using radioactive technitium 99, which enables a computer-assisted camera to take pictures (scintiscan) of the heart muscle after an attack. It can show the doctor where the damage is and, in some instances, give specific measurements of the size of the infarction. By using it with another radio-nucleide technique — thallium 201 — your doctor can be told if the damage that shows up is old or new. The most recent accomplishment was reported in the April 14, 1977 *New England Journal of Medicine* by a team of NHLBI researchers who have increased the sensitivity of the new techniques so that they can now get pictures of your heart muscle at work during exercise as well as at rest. Not surprisingly, the new procedures are very expensive, not too readily available, and still subject to improvement. The future for better diagnosis, however, is unquestionably brighter.

And while answers to old questions are still not all in, new questions continue to arise with regularity. Why do

---

[52] Developed at the University of Texas Health Science Center, Dallas.

Americans with the same risk factors as Europeans have twice the rate of CHD? asks University of Minnesota's Dr. Ancel Keys, principal architect of worldwide studies. Why do victims of Down syndrome (mongolism — a genetic disorder involving an extra chromosome) die relatively free of atherosclerosis, in contrast to other patients in the same institution? asks a July 1977 report in the *British Medical Journal*. Why do persons who drink hard water have a lower rate of CHD than soft-water imbibers? Exactly what role do trace elements, including zinc and copper, play?

For David Miller, the answers to his most pressing questions are in. From the beginning, he was not afraid to acknowledge to himself that he was afraid. He faced another crisis when fear of dying gave way to fear of living, until he was persuaded that his was not to be a life of stringent sacrifice (the worst sacrifice is cigarettes) and restriction. He knows his limitations, but they in no way interfere with his life. His doctor also knows *him*, and following Osler's advice that "it is often more important to know what sort of patient has a disease than to know what sort of disease the patient has," he allowed him out of bed early while still in the hospital, and back to work not long after the Millers' visit that opens this chapter.

Most important is David's reassurance of the umbrella of protection under which his children will now mature. With the alert of their vulnerability sounded and their willingness to listen, the outlook for their future is good for many decades ahead.

3

High Blood Pressure — A Family Affair

There was a murmur of surprise in the crowded meeting hall when noted pediatrician Dr. Sidney Blumenthal[1] told the New York Academy of Sciences 1977 Hypertension Conference[2] that "identification in early life of the genetically susceptible individuals is a problem of new and high priority." The medical community was not yet thinking of high blood pressure as a genetic disorder, nor were all pediatricians looking for signs of it in their young patients.

Yet there is hardly a person touched by the problem in adulthood who does not know that it is often a family affair and that you are at greater risk if either or both of your parents or any brothers or sisters are affected. "Family aggregation of blood pressure is already well established in early childhood," says Dr. Edward A. Kass, whose Harvard Medical School team[3] brought their spe-

---

[1] Director, Division of Heart and Vascular Diseases, NHLBI.
[2] "On Mild Hypertension: To Treat or Not To Treat."
[3] Including Drs. Stephen H. Zinner and Paul S. Levy.

cially designed blood-pressure recorders into the homes of 190 families with a total of 721 children ranging in age from 2 to 14. Not only did the children as young as 2 already register blood pressures close to those of other members of the family, but when the same children were looked at four, six and eight years later, they were still on the same track. Those who were higher initially were still high compared to their peers, and those who started out low were still on a low track.

Researchers now know that the family resemblance shows up as early as one month after birth, with the blood pressure correlation between mother and child already present at four days. As the months and years go by, correlation within the family (especially among brothers and sisters) increases, reaching a peak by the age of 20, reflecting the impact of a shared environment on the genes that control blood pressure.

How much is heredity and how much is environment? "Almost 60 percent is genetic," says Dr. Manning Feinleib, a member of an NHLBI team that has examined 514 adult male twin sets at five centers across the country. The heredity factor may be even stronger in very young children, as was shown by University of Montreal's Dr. Paul Biron, who studied 274 families with at least one adopted child in addition to natural children. Correlation between the adopted children and the rest of the family was *zero* when the adopted child was less than a year old at adoption and not yet seven at the time of the study. After the age of seven, the adoptee began to show a "slight but nonsignificant correlation" with the rest of the family. The natural children, on the other hand, had the same family resemblance in blood pressure as the Boston children, Miami children, Maori children in New Zealand, and Polynesian children on Tokelau Island.

The fact is that genes for hypertension are to be found

in all countries and among all peoples — including members of primitive tribes in New Guinea and high up in the Andes; residents of affluent suburbs, crowded city ghettos and rural regions; fishermen in coast villages of Japan and nomads in the deserts of Iran. Yet there are populations free of the disease, and others with alarming prevalence. Whether the defective genes will be expressed is directly related to how much you weigh, how you cope with stress, how much you drink and smoke, how you control fertility, and how you season your food. These are among the environmental forces already operating in the first two decades of life, underscoring the urgency for early identification of the 15 to 20 percent of the non-blacks and the 30 percent of the black population in the United States genetically susceptible to hypertension.

The focus on the young is a new phase of the National High Blood Pressure Education Program[4] launched in the spring of 1972. Rarely has a major public health campaign faced as formidable a challenge. The target — high blood pressure (hypertension) — was described at the time by cardiac epidemiologist Dr. Jeremiah Stamler as "one of the most important, if not *the* most important, affliction producing premature disability and death in our adult population." Summarizing the immediate problem as "one of simple arithmetic — ½ × ½ × ½," Dr. Stamler stressed that of the estimated 23 million Americans affected, half were unaware. Among the half who knew, half were paying no attention to it, while among the remaining half

[4] Spearheaded by then Secretary of HEW Elliot Richardson and coordinated by the NHLI of the NIH, under the overall direction of Dr. Theodore Cooper, the National High Blood Pressure Education Program enlisted the enthusiastic participation of more than fifty government, professional and voluntary organizations and industry, and now involves 150 organizations.

under treatment, no more than half were adequately controlled.

The grim statistics are that hypertension is the leading
cause of death among United States blacks (for every black
who will die of sickle cell disease, a hundred will die of
hypertension) and a major cause of death among non-
blacks. The irony is that among all the life-threatening
disorders to which we are prone, high blood pressure
ranks highest in the ease and convenience with which it
can be detected, and the effectiveness, safety and speed
with which it can be controlled.

Nevertheless, despite numerous warnings dating back
to the 1950s, the majority of the American public did not
perceive hypertension as a threat to *their* health — not
surprisingly, since it can be present for many years before
a single symptom surfaces. Frequently, that first symptom
is a catastrophic stroke, heart attack or kidney failure.
Moreover, many consider it primarily a disease of the
elderly. Not so. Approximately 85 percent of all strokes
— most of them preventable — happen to persons in their
fifties and sixties, and it has been estimated that 40 to 70
percent of all deaths in middle age are associated with the
ravages of untreated and uncontrolled high blood pressure. Paralleling the education of the general public was a
program aimed at physicians, many of whom were not yet
thinking in terms of instituting lifetime surveillance and
treatment of a patient who came to the office without a
complaint.

"I now tell all my patients in their forties," says one of
our favorite practitioners, "that if they were as concerned
about their blood pressure as they are about a receding
hair line or a few lines under the eyes, they could add
fifteen years to their lives."

By 1977, hypertension was no longer a "silent
epidemic." At least three million men and women, hereto-

fore unaware, had found out — at their doctors' offices and at community-based screening programs designed to reach them where they worked, studied, relaxed and worshipped. Marian received her warning at a shopping center; Russell learned about himself at a church bazaar; Sylvia was identified at the employees' health clinic; Josh came to his doctor, rolled up his sleeve, and explained: "My brother is in the hospital with a stroke," reflecting the increasing public awareness that hypertension is a family affair.

If Marian, Russell, Sylvia and Josh (all under 50) are typical, the odds are that today only Sylvia (possibly) and Josh (probably) are following through on the initial alert. While the numbers under good control have doubled, close to 50 percent still remained under poor or no control in 1977. And so, with no cure in sight in the foreseeable future and widespread control elusive in the present, the new look is at prevention.

Until recent years, efforts to convince both the public and the profession that hypertension and its consequences could be prevented would have been based on meager evidence. Even today, with much of the mystery that still surrounds the underlying causes and mechanisms of the disease, prevention sounds like an extravagant promise. However, enough is known about it to view with confidence both the long-range goals with children and the more immediate problems of the 23 million adults, of whom more than 20 million are still mild but are at greater risk for trouble ahead than their normotensive counterparts.

High blood pressure is a unique disorder — easier to diagnose, treat and control than to define. Indeed, so complex are the mechanisms that determine the pressure under which your blood is pumped into and flows through your arteries that it is a triumph of nature that the narrow range associated with good health is maintained most of the time by most of the population. Two numbers describe blood pressure. The higher number — systolic — the pressure generated during contraction of the heart muscle as it pumps; the lower number — diastolic — when the heart is relaxed immediately before the next contraction. 120/80 is considered optimal in adults, but there is no sharp dividing line between normal and abnormal.[5] During the course of the day and night, your blood pressure fluctuates with heat and cold, pain and pleasure; it goes up when you are excited and is low when you are asleep. Only when the elevated pressure is sustained over a period of time, with repeated readings, will your doctor diagnose hypertension.

Two deceptively simple criteria determine your blood pressure — the volume of blood circulating in your vessels, and the resistance of the walls of the arteries as the blood flows through. The controlling mechanisms, however, are enormously complex, and hypertension can best be defined as the disorder of one or more of these mechanisms. Involved are cardiac output; secretion of the adrenal glands; enzymes from the kidneys; size and structure of the blood vessels; viscosity, volume and electrolyte content of circulating blood; thermostat-like structures (baro-receptors) that send signals to the central nervous system. "The brain may well serve as the focal point for setting into play the many diverse elements," says Dr. Donald J. Reis, Director, Laboratory of Neurobiology, New York Hospital–Cornell Medical Center.

The 1970 landmark Veterans Administration Study was to move the treatment of hypertension away from the crisis medicine of strokes, heart failure and kidney deterioration, to the medicine of prevention. Furthermore, it was to destroy the myth that asymptomatic moderate

---

[5] Up to 90 diastolic is acceptable; 90–95 borderline.

elevations of blood pressure can be ignored. The patients in the study were men with modest blood pressure increases — 90 to 114 diastolic. They were treated for a period of five years with carefully selected choices and combinations of drugs developed since the 1950s but never before utilized so precisely. Not only did treatment reduce or eliminate completely familiar complications of strokes and congestive heart failure; it also prevented blood pressure from rising to more dangerous levels.

"The report did not create much of a stir when it was published in the widely read *Journal of the American Medical Association* on August 17, 1970," Georgetown University's Dr. Edward T. Freis, chairman of the study, told me not too long ago. "It was not until several months later at an epidemiology meeting that its full impact came through."[6] The action after that was swift and bold.

Two of the figures who played a key role in initiating the national program had firsthand knowledge of the hazards of the disorder. HEW Secretary Richardson saw his father die at the age of 46 from complications of hypertension, and so he was particularly receptive when the woman who is credited as the "catalyst" in the campaign came to him, armed with the facts on the dismal status of hypertension control and the bright prospects held out by the Freis report. Mary Lasker's contributions to promoting the health of the nation had already been widely recognized; she holds a number of honorary degrees from a number of medical schools.[7] This time, the public health problem of hypertension had a personal meaning for her — both her mother and father had died of strokes. The American Heart Association, mindful that

---

[6] *JAMA*. The implications were so far-reaching that only one year later Dr. Freis was honored with the coveted Lasker Award.

[7] *Who's Who in Health Care*, first edition, 1977. It is also noteworthy that the prestigious annual Lasker Awards for medical research have, on more than one occasion, been a prelude to a Nobel prize.

hypertension was the #1 contributor to atherosclerosis — the forerunner of coronary heart disease — agreed that an attack on high blood pressure was a #1 priority.

For the hypertensives who got the message, the benefits in the first few years of the program have been as impressive as for the veterans. Strokes have plummeted a dramatic 14 percent, and coronary heart disease declined by 7 percent. Although it was not possible in 1970 to document the connection, in 1977 reports from Britain revealed that treated hypertensives suffered half as many CHD deaths as the untreated. "Today," says Dr. Marvin Moser, senior consultant to the national program, "in well-treated patients, heart failure has been virtually eliminated as a complication of hypertension; deaths from malignant [i.e., rapidly accelerating] hypertension have declined almost 80 percent during 1958–74; massive brain hemorrhages are rare." Moreover, controlling hypertension is a factor in the fall in the death rate from all cardiovascular diseases (which is almost twice as large as from other disorders) since 1950.

With the consensus among researchers today that high blood pressure has its roots early in life, there is a growing awareness that most of the victims in their fifties could have been identified in their twenties and thirties. Moreover, brothers and sisters of affected persons tend to have higher-than-normal blood pressure even if they are not yet in the abnormal range.

"We don't share much in looks, attitudes or lifestyle," says Lucy about her family, "but our kinship shows up in our blood pressure." Now 54, Lucy first found out that there might be a problem almost twenty-five years ago, the night after the birth of her third child. "The resident was buzzing around at 1 A.M.," she recalls, "and I panicked. I felt fine, the baby was perfect, but my blood pressure had shot all the way up. By the next day, it was back to normal."

Her doctor was not alarmed, but he advised her to "get rid of the salt shaker" while he continued to watch her periodically for the next decade and a half, as the numbers fluctuated, always hovering at the upper limits of normal.

For the last ten years, since that limit was exceeded, she has been on therapy. Lucy is an unusually motivated and disciplined person — she is determined to give herself the best opportunity for the added longevity therapy promises. Memories of her father's death from coronary heart disease following ten years of hypertension, and her mother's premature death of advanced arteriosclerosis after many years of diabetes, are a sobering factor. How is she today, twenty-five years after the first alert? Still a slim, dynamic community leader, a golfer, a grandmother, and still free of a single symptom of what she accepts as a lifelong disease with which she can deal.

Lucy's brother was 23 when he found out that his blood pressure was "not anything to worry about" but too high to qualify for a pilot's license at the time. He, too, was watched by the doctor for ten years, kept his weight down, cut his smoking to only one cigarette after each meal, and is, like Lucy, on a minimal therapeutic program which keeps his moderate and still asymptomatic hypertension under control.

Their younger brother, on the other hand, is having a harder time bringing down his newly discovered high blood pressure at 48. Never having identified himself with the family vulnerability, he first learned about it when he complained to his doctor of unexplained weariness and headaches. Now he has a problem losing his twenty-five extra pounds and confronting a day with fewer than the pack of cigarettes to which he is accustomed; and the fatigue that brought him to the doctor is not yet gone. Could his condition have been identified earlier? Probably. Would control have been easier? Almost certainly.

Until not too long ago, "essential" or primary hypertension in the young was considered rare. While more than 90 percent of adults fall into that category, inexplicably labeled "essential" because the cause is unknown, it was believed that in the young, hypertension was more likely to be secondary to another condition[8] and usually correctable by surgery. Today, however, primary hypertension is rapidly being established as a disease of childhood and adolescence as well.

How early can blood pressures be recorded? A research team from the University of Florida School of Medicine headed by Dr. Mary Jane Jesse (now at the NHLBI) reported in 1976 that they were able to obtain routine measurements with ease in a group of 249 infants by using an ultrasonic device. Blood pressures were taken at two days, one month, three months, six months and one year. In a blood-pressure survey of 391 newborns, London's Dr. M. de Swiet and his colleagues[9] found that those whose pressure was higher at four to six days continued to show a relatively higher pressure at five to seven weeks. "If this trend continues with age, it would suggest that the tendency to develop hypertension may already be demonstrable at the age of four to six days," they conclude.

Taking blood pressures in the nursery is not yet (and may never be) a routine procedure. But when the 1977 recommendation of the NHLBI Task Force on Blood Pressure Control in Children[10] is adopted, it will soon be routine in every pediatrician's office and clinic. Convened in response to repeated inquiries from pediatricians, school health personnel, community health planners, and

---

[8] Secondary to kidney disease, adrenal tumor, thyroid disorder, defect of the aorta, and other disorders.

[9] *British Medical Journal*, July 3, 1976.

[10] *Pediatrics*, May 1977.

clinical investigators, the Task Force, chaired by Dr. Blumenthal, came up with guidelines for action now. Blood-pressure measurements should start routinely at the age of 3, and be recorded as part of the health record, along with height and weight. Included in the guidelines is a statement of the importance of family history of primary hypertension, particularly in parents and grandparents who are less than 50 years old.

If blood-pressure readings in young children are to be a meaningful index of their future risk, they must be taken under standard conditions from year to year, says Cornell University Professor of Pediatrics, Dr. Maria New. "I take it three times at each visit," she told me, "and record what the child was doing at the time — sitting, lying down, protesting, crying. That's important for the next examination." Finally, Dr. New emphasizes that "blood pressure must be a part of the *whole health care* of the growing child."

While hypertension in a very young child is more likely to be secondary, a pattern of increasing prevalence of primary (essential) hypertension among adolescents is emerging from recent studies of school-age and college students across the country — New York City; Davis, California; Dallas, Texas; Muscatine, Iowa; Cincinnati, Ohio; Richmond, Virginia; and Evans County, Georgia. Many have a family history of hypertension, and a disproportionate number are obese. It is still not known how many will go on to develop hypertension in later life, but a special concern is the large number of young persons with a fluctuating or casual high pressure. In adults, it is a danger sign, and a number of studies suggest that it may be so in youth as well.

"Teenagers and young adults who 'spike' in high school or college, pre-employment or Armed Forces physicals," maintains Dr. Stamler, "are not people to be ignored in terms of significance of such 'spikes' for future disease."

Even a single casual high-blood-pressure reading in youth can identify a person as a future risk. When the University of California epidemiologist Dr. Ralph Paffenbarger, Jr., studied the college records of 45,000 Harvard and University of Pennsylvania alumni, he tracked down the early history of 590 who died of CHD in their middle forties. *All had had higher blood pressures than their surviving classmates when they entered college as teenagers.*

Studies of Chicago gas company workers, New York insurance employees, and others whose health histories were followed for twenty years or more told the same story. Those with elevated blood pressure early in life suffered more heart disease and more sustained hypertension later in life. For some young hypertensives, the danger is not far off. In Evans County, Georgia (the stroke belt of the United States), a group of thirty adolescents with elevated blood pressures were watched for seven years. During that time, almost one-third developed sustained hypertension and half a dozen underwent vascular changes. By contrast, the young people with normal pressure at the beginning of the study were still normal. For at least two young women (both black), discovery in adolescence was not early enough. They died of cerebral hemorrhage.

"Find the environmental factors," warns Dr. Kass, "and remove them as early as possible." One of the most common factors, in both adolescents and adults, is obesity. (In the Evans County Study, it was two to three times as common in the high-blood-pressure group as in the normal group.) And frequently, just losing weight and keeping it down with diet and exercise is the only treatment needed to reduce blood pressure in the adolescent or adult with mild hypertension.

Cutting out cigarettes *cuts risk by 50 percent*. While there has been a significant decline among middle-aged men (who are very vulnerable), women have cut down far less. Teenagers, on the other hand, are smoking in larger numbers than ever. Reports about the contribution of the Pill to heart disease and hypertension continue to confirm earlier warnings both in the United States and abroad. Britain's Royal College of General Practitioners finds four times as many strokes in users (mostly older women). A June 1977 Kaiser-Permanente study of 13,350 women recorded a slight but statistically significant rise in blood pressure in users. Alcohol — while one or two drinks a day may relax you, more will push your pressure up.

New insights into the role of stress are emerging from the laboratories of neurophysiologists, pharmacologists and psychologists. And to help you cope, there is a renewed look at old techniques — including meditation and relaxation.

Occupying center stage among risk factors is one of the most popular and irresistible food additives — salt.[11] For those unable to handle it properly, it is the most devastating contributor to trigger lifelong disease. It is also the *one common factor* in the environment that (1) acts early in life; (2) acts on the genetically susceptible all over the world, with different ethnic origins and living in settings that range from searing deserts to frozen Arctic wastes. Indeed, so certain are some of the leading experts on the subject that a growing number are now confidently promising that it would be possible to eradicate hypertension as a major public health problem in one generation.

□ Dr. Edward Freis: "The evidence is very good, if not

---

[11] It is the sodium in table salt (sodium chloride) that contributes to hypertension. Other sources of sodium to be watched are monosodium glutamate (MSG) and sodium nitrite in bacon, salami and other processed foods.

conclusive, that reduction of added salt in the diet to below two grams a day (9/10 teaspoon) would result in the prevention of essential hypertension and its disappearance as a major public health problem."

☐ Dr. Lot B. Page, Chief of Medicine, Newton-Wellesley Hospital, Massachusetts, and Professor, Tufts–New England Medical Center: "Low salt in infancy may eliminate the *whole* public health problem. . . . I would be willing to settle for 50 percent if it can be done in a generation."

☐ University of Mississippi's Dr. Herbert Langford: "Abolish salt, abolish high blood pressure."

But even the strongest proponents of removing salt from the human experience are realistic enough to acknowledge that what is possible today is far from probable. "As far back as ancient times," says Dr. Page, "salt has been associated with good words — *sal*vation, *sal*utory, etc." Furthermore, a taste for salt develops almost from the earliest exposure to it, and for many, the preference develops into a craving not easily denied.

The importance of salt as an initiator of hypertension was recognized by investigators as early as the turn of the century, but it was not until recent decades that convincing (albeit to some extent circumstantial) evidence has been emerging. From studies of primitive peoples in various parts of the world, scientists now know that not only are there whole populations in whom hypertension is absent, but the long-held belief that it is normal for blood pressure to rise with age is not true. On the contrary, they reach advanced age with no higher pressures than their grandchildren. These are, for the most part, the people who have not yet been assimilated into western civilization and habits, and have not yet acquired a taste for salt.

They include Greenland Eskimos, Polynesians from isolated islands in the Pacific, Easter Islanders, nomadic tribes in Kenya, Congo pygmies and inhabitants of rural Uganda.

Nor do they suffer from their salt intake if it is limited to the content of their natural foods. The bushmen of the African Kalahari Desert were in excellent physical condition when they were studied in 1960, and were able to run for long distances in pursuit of game. Especially impressive are the Masai of Africa, traditionally a war-like people of splendid physique. "Low-blood-pressure, low-salt populations who have lived for many generations in desert, Arctic, and jungle environments use less than two grams (9/10 of a teaspoon) per day . . . ," says Dr. Page. "In spite of heavy labor, sweating and breast feeding [nursing mothers are said to need more salt], they have never suffered from any deficiency state."

Dr. Page was a member of the Harvard expedition that studied eight racially homogeneous Solomon Islands populations from 1966 to 1972. All lived traditional lives in rural areas, but in a setting where western influence was beginning to be felt. Members of the three groups most acculturated showed a rise in blood pressure with age, so familiar to us. Among the five other groups, still relatively unchanged, low pressures persisted. When the impact of *all* the new factors was assessed, ". . . a single dietary item, salt, was more related to the blood-pressure trends than were any of the other factors singly or collectively," Dr. Page asserts.

Even primitive people are not genetically immune to hypertension; the appropriate trigger — salt — is capable of unmasking the heretofore silent gene. So it was with the Lau tribe in the Solomons, who ranked third in acculturation but were far above all the other tribes in elevated blood pressure. They had a long-established custom of boiling vegetables in sea water and, consequently, the highest salt intake of all eight groups. Dr. Page's most recent expedition took him to the deserts of southern Iran, where he and his colleagues lived with the nomadic Gashgai tribe. They are known to consume large amounts of salt, and, like the Lau of the Solomon Islands thousands of miles away, are relatively untouched by western habits. Gashgai blood pressures are strikingly high.

Nowhere in the industrialized world is the effect of salt felt as strongly as in Japan, where stroke is the leading cause of death. The highest incidence of hypertension is found in northeastern Japan, where the average daily salt intake in fish, soup and pickles is almost one ounce, and where hypertension among persons in their forties is 30 to 40 percent more common than anywhere else. Almost as high is hypertension among Newfoundland fishermen, who also consume inordinate quantities of salt.

The first concrete evidence of the link between heredity, hypertension and salt came from the pioneering studies initiated in the 1950s by Dr. Lewis K. Dahl, Medical Research Center, Brookhaven National Laboratories. In a survey of a large number of adults whom he divided into three groups according to salt intake, he found that among the high group, over 10 percent had hypertension; among the average group, 7 percent; and among the low salt users, less than 1 percent. Finding out why from animal experiments in the laboratory posed a problem, for animals do *not* spontaneously develop hypertension. By feeding equal amounts of salt to randomly selected rats, he found that some were sensitive (S) and developed high blood pressure, while the rest, resistant to salt feeding, retained normal pressures (R). Inbreeding offspring within each group yielded two distinct strains — one of animals whose heredity put them at high risk for hypertension when fed large amounts of salt, and others of the same species who were endowed with genes that permitted them to do well, even in the presence of salt.

In the early 1960s, Dr. Dahl and his colleagues reported

(1) persuasive proof that in rats, the predisposition to high blood pressure is inherited; (2) the earlier in life the susceptible rat is exposed to salt, the more severe is the hypertension; (3) the impact of the dietary salt is not apparent at once — it takes time for the hypertension to develop; (4) once it sets in, it is virtually irreversible — only drastic reduction in salt intake can modify the disease; (5) even the susceptible rats *do not develop hypertension if they are kept away from salt*. Clearly, their genes are *not* their destiny.

These were startling observations indeed, and considering the implications for human health (much of rat physiology is close to human physiology), Dr. Dahl's discoveries created a relatively minor stir. To be sure, he found recognition and support from a few researchers — notably Harvard nutritionist Dr. Jean Mayer (now President of Tufts University) — but almost a decade was to pass before Dr. Dahl's warnings were to be applied. A major concern of his at the time was the copious amount of salt in commercially prepared baby foods. The salt was probably harmless to 80 percent of all infants, but it was a trigger that could set the genetically susceptible infant on the way to a lifetime of high blood pressure. In experiments with thirty different varieties of baby foods — all of which showed a salt content greatly in excess of that of the unprocessed meats and vegetables from which they were made — the researchers found that susceptible rats developed hypertension within four months.

"The evidence that dietary salt will induce permanent experimental hypertension in rats is direct, quantitative, and unequivocal. . . . At this juncture, it is *necessary* to transfer such experimental results to man," Dr. Dahl wrote in *Nutrition Reviews*, April 1968. To the skeptics who argued that maybe more experiments were needed to determine the salt requirements of infants, Dr. Dahl replied, "Such an experiment has already been carried out during tens of thousands of years when human milk was used as the only source of food and fluid well into the first year of life." Noting that human breast milk is *very low in salt* (with one-third the amount of salt in cow's milk), he commented that "if breast-fed infants develop a sodium [salt] deficiency, I am unaware of it."

It was not until 1974 that some manufacturers reduced salt content in some of their baby foods. In the spring of 1977, two of the major companies — Heinz and Beechnut — announced that they were eliminating added salt from all of their baby-food products. Gerber said it plans to offer some varieties either with or without added salt.

Why the emphasis on restricting salt from infancy on? The exact mechanism of salt's role is not yet known, but evidence points strongly to a defect in the kidney's ability to get rid of it in the genetically susceptible. What is known is that it sets in motion an upward trend that eventually is irreversible, and can no longer be corrected by just throwing away the salt shaker. Since most processed foods and snacks are liberally seasoned with salt, you could not easily escape it even within the confines of your own home. Actually, all you need is 1/10 of a teaspoon of salt (200 milligrams sodium) a day. A total of one teaspoon, including the sodium content naturally present in meat, fish, milk and vegetables, can be handled by most persons. But the typical American diet contains many times that amount, with teenagers the worst offenders. At least 50 percent of all items on their diet are high in salt.

Dr. Freis believes that there is enough evidence today to say that all of us, compared to our early ancestors, suffer from a salt-induced chronically expanded blood volume, a condition "nature did not intend for us to handle." For those at risk because of a hereditary defect in kidney function or other mechanisms for handling the increased load, the higher than normal pressure becomes a permanent problem, reversible only by bringing the fluid volume down with diuretics or an unacceptable virtually salt-free diet.

Rivaling the role of salt in hypertension is the role of stress, familiar to every doctor who treats hypertension and to most persons who suffer from it. Josh's pressure goes up every time he walks into a sales conference, even when he is about to report on a lucrative new account. Marian does well both in her home and at her job, but she can't take crowds. Lucy's pressure sometimes still spikes when she mounts a platform to champion an unpopular school budget in a community increasingly resistant to rising taxes. Numerous studies among workers threatened by loss of jobs tell the same story — a significant rise in blood pressure.

Nature provides a mechanism to deal with stress — the fight or flight response[12] to a threat, a fear, an emergency. The message goes out from the brain and other parts of the nervous system to target organs in the first line of defense. Heart output and respiration increase, sending more blood to your muscles (as much as 300 to 400 percent more). Adrenalin and related chemicals (catecholamines) are poured into your circulation, and your blood pressure goes up. When you can escape the threat (flight) or resist and overcome it (fight) and it is not a frequent and repetitive occurrence, the mechanism serves you well. When, however, you keep eliciting the response in situations doomed to failure, the repeated transient elevations in blood pressure can, in a vulnerable individual, ultimately become a severe sustained hypertension.

A striking example of what continued frustration and helplessness can bring about is what happened to the head

[12] Described at the turn of the century by Harvard physiologist Walter Cannon.

of a family of baboons studied by Russian researchers. The family consisted of one male living in a large cage with his own harem of females. He was removed and placed in a nearby cage alone, while a new male took over his cage and his females. For one year the displaced male watched the new relationship, and he ended up with severe hypertension, a heart attack and a stroke!

New evidence on the combination of heredity, stress and salt in hypertension comes from Dr. Dahl's laboratories, where the research he pioneered on genetically susceptible rats is continuing despite his untimely death in 1975. Drs. Richard Friedman and Junichi Iwai put rats in a cage for a few hours a day, where they learned that to obtain food they had to press a lever. The problem was that the same lever could also yield an electric shock. Since both food and shock delivery schedules varied, there was no way for the rat to predict if pressing the lever would give him food, a shock, neither or both. Exposed for twenty-six weeks to the stress of the chronic food-shock conflict, the genetically susceptible animals developed a persistently elevated blood pressure, even on a very low sodium diet. By contrast, genetically resistant rats suffered no such damage when confronted with the same conflict.[13]

Nature also provides a mechanism to diminish the consequences of the failed "flight or fight" response — relaxation and meditation. "Techniques for eliciting it have existed for centuries in virtually every culture, often in a

[13] Later experiments showed that more salt added to the diet produced an even greater elevation of blood pressure in the susceptible. In another phase of the study, the scientists found that large amounts of salt still surpass stress as a threat. Together, they are a sinister combination. *Proceedings of the Society for Experimental Biology and Medicine*, July 1977.

religious context," says Harvard University's Dr. Herbert Benson, author of the best-seller *The Relaxation Response*. The Hindus practiced it as far back as the sixth century B.C.; the Jews used it at the time of the Second Temple in the first century B.C. and the Christians in the eleventh and twelfth centuries; and the fourteenth-century Jewish cabbalists included it in their ritual.

When they sat quietly and comfortably, not allowing themselves to be distracted as they kept repeating a simple word or short phrase, they found spiritual fulfillment. An added bonus of which they may have been unaware was a slowing down of the physiological processes that push blood pressure (and other disorders) up.

"It is a natural process," says Dr. Benson, "and we have stopped using it because our society does not condone its use." It is, however, staging a comeback in a secular setting, among thousands who are just seeking relief from the stresses of daily living. And now for the first time, the techniques, including relaxation, meditation and biofeedback, are being talked about seriously by researchers as a new way to (1) diminish the physiological events generated by stress; (2) prevent development of hypertension in persons in the high normal range or at risk for a variety of reasons; and (3) provide help for the 20 million Americans with mild or borderline elevation, for whom a lifelong dependence on drugs may not yet be the answer.[14]

Down through the centuries, while yogis and other practitioners of meditation and relaxation continued to demonstrate that they could control their heartbeat, raise and lower temperature, direct blood flow to and away

---

[14] The Hypertension Detection Follow-Up Program, NHLBI, estimates that 70 percent of all adult hypertensives have diastolic pressures of 90-105, for whom drug therapy is not routinely recommended.

from selected parts of the body, conventional thinking dictated skepticism that these "involuntary" functions could be changed at will. Scientists have long been locked into the traditional view of two types of behavior. The voluntary system, which allows you to make decisions and to act appropriately, is based on learning and experience and is controlled by the "thinking" region of the brain. And so it is that you consciously rush forward to meet a friend, or quickly cross the street to avoid someone you prefer not to greet. By contrast, the traditional view held that the autonomic nervous system — the regions of the brain and the nerves that control breathing, heart rate, digestion, blushing, sweating, hormone release, blood pressure — could not be under voluntary control because, as one skeptical scientist recalls, "the autonomic nervous system was 'stupid' " — you could not teach it to order your heart to pump more slowly and your blood pressure to fall.[15]

The myth was dispelled in 1970 by the skeptical scientist, Rockefeller University's Dr. Neal Miller and a team that included Dr. Leo V. Di Cara and a number of open-minded graduate students, when they introduced a new concept to the everyday vocabulary — biofeedback. "Most people are unaware of their blood pressure," Dr. Miller points out, "and expecting them to control it voluntarily is like expecting a blindfolded basketball player to learn to shoot fouls. Since he does not know where the ball goes, he is doomed to failure." The researchers removed the "blindfold" with a device that gives a continuous reading of the patient's blood pressure. When the pressure goes down. the electronic sensing device amplifies the

---

[15] Nobel Laureate Dr. Roger Guillemin believes there is a third nervous system based on the activities of the newly discovered brain hormones. See Chapter 4.

signal and feeds it back instantaneously. The patient now knows to continue whatever it was he was doing or thinking when the pressure fell. To reinforce progress, biofeedback often includes a system of rewards. Where the reward is freedom from pain or freedom from disease, biofeedback has produced demonstrable success in lowering blood pressure, controlling irregular heartbeat, migraine headaches, convulsive seizures, severely painful muscle spasms, and anal incontinence.

"The changes, however," Dr. Miller explained in a 1977 interview, "are not clinically significant, and too short-lived to be considered therapeutic. With more rigorous and well designed studies, in five years the field may be shaken down, and we will then know how and where it can be of real use."[16]

Biofeedback has already been of "real use" — by removing the blindfold from researchers who still thought the autonomic system could not be modified. They then took a new look at an old therapy that is simpler than biofeedback and not dependent on instruments — relaxation. Indeed, before the development of effective anti-hypertensive drugs in the 1950s, the relaxation achieved by the doctor's reassurance and understanding (often supplemented with sedatives) or by psychotherapy was the only course of treatment. Psychotherapy was too costly for many and not successful for others. The casual and sporadic degree of relaxation attributed to reassurance was too often of too short duration.

Meanwhile, yoga and meditation, long practiced in

[16] The Rockefeller group scored one spectacular success with a young man paralyzed by a spinal injury, who suffered an abnormally low blood pressure in an upright position. He was so determined to learn to walk on crutches that with biofeedback he learned to bring his pressure up. He has walked with crutches and braces for the last four years.

eastern religions, commanded a new interest, both among investigators, who confirmed with physical measurements the technique's success in self-regulation of heart, temperature, etc., and among an increasing segment of the general population. Among the most popular is T.M. — Transcendental Meditation — introduced to the western world in the late 1950s by the Hindu monk Maharishi Mahesa Yogi[17] and now claiming half a million adherents in the United States. Relaxation and meditation are achieved with the help of a special personal two-syllable word called a mantra, a device that diverts concentration from external reality into oneself and a state of heightened consciousness. Its benefits are enthusiastically acclaimed by sports personalities, business executives, members of the armed forces, and whole families who train for it at the same time.

Much of the credit for acceptance of relaxation as a therapeutic measure goes to Dr. Benson, who was the first to adapt meditation to a simple, non-cult procedure easily mastered by and more acceptable to persons who are uncomfortable with the exotic aspects of T.M. The first technique specifically developed to control high blood pressure, it is also proving useful in controlling one form of irregular heartbeat, and in reducing anxiety in preoperative patients. It may even be helpful in correcting skin disorders.

We first met Dr. Benson at an American Association for the Advancement of Science research conference in New England in the early 1970s, not too long after he had published his first papers on the physiology of

[17] Although disciples of the Maharishi deny a religious component, a Federal District Judge in New Jersey ruled in the fall of 1977 that T.M. could not be taught in four public school districts where it had once been offered under a government grant.

meditation[18] and the lower blood pressures in hypertensives who meditate.[19] "The idea for the study came from my students. It was a time when 'relevancy' was on everyone's mind, and we agreed that looking into meditation and hypertension was relevant."

Relaxation, Dr. Benson reported, was accompanied by measurable changes — a fall in blood pressure and a decrease in metabolism, oxygen consumption, carbon dioxide elimination, and blood lactate (a substance associated with the anxiety state).

The usefulness of the relaxation response in the treatment of hypertension has since been confirmed by Dr. Benson and his co-workers in ongoing studies, as well as by investigators across the country and abroad. At the University of California, San Diego, Drs. Richard A. Stone and James de Leo used Buddhist meditation exercises in a six-month trial with a group of young relatively mild hypertensives. Significant decreases in blood pressure were achieved in more than 50 percent. Their study also included biochemical measurements documenting the role of the central nervous system in hypertension. In London, Drs. C. Patel and W. North used a combination of yoga, biofeedback and instruction on the basic facts of hypertension, and noted a substantial fall in blood pressure on follow-up. Patients on drug therapy were able to get along on substantially smaller doses. No matter whether the relaxation is through T.M., yoga, or relaxation response, the results are in agreement.

Physicians who not too long ago scoffed at the potential of the ancient therapy are now reading about it with increasing frequency in the most prestigious medical publications. Dr. Benson put the subject into perspective in

[18] *Scientific American*, February 1972.
[19] *Circulation*, Vol. 46, Suppl. II, 1972.

the May 19, 1977 *New England Journal of Medicine*. It will never take the place of drug therapy in the moderate or severe hypertensive, he said, but for the mild and border-line, it can supplant drugs completely. For those on drugs it can reduce the doses, and for the still normal at risk, it can be a preventive measure.[20] Moreover, says Dr. Ben-son, "it is a simple natural phenomenon; does not require complex equipment for monitoring . . . or involve the expense and side effects of drugs."

University of Pittsburgh's Dr. Alvin P. Shapiro, a member of an NHLBI ad hoc committee set up to review non-drug treatment of hypertension, reported to the New York Academy of Sciences Hypertension Conference on the current status. "Biofeedback and relaxation techniques are of special interest," he said, describing them as "fresh approaches . . . stimulating investigators and clinicians to look more seriously at the behavioral influences of hyper-tension . . . to develop methods to utilize them more precisely . . . and to teach them to health professionals and to patients."[21] At the same time, Dr. Shapiro cautions that the present studies, while they are encouraging, rep-resent a relatively small number of individuals who have been tested by rigorous scientific standards.

Finally, a guarded but unmistakable optimism is ex-pressed editorially by:

☐ *New England Journal of Medicine*, January 8, 1976:
   "If autonomic functions can be modified and trained in man, and if such training is effective, and more impor-tant, if it can be maintained, . . . sustained hyperten-sion may be prevented. . . . It is worthy of further evaluation."

☐ *British Medical Journal*, June 12, 1976:

---

[20] *American Journal of Public Health*, October 1977.
[21] Also reported in the May 1977 *Annals of Internal Medicine*.

". . . these studies raise the possibility that control of blood pressure in some well-motivated patients with mild hypertension may be possible without drugs. . . . There is no immediate suggestion that physicians' efforts to lower blood pressure with drugs will be replaced wholesale by relaxation techniques. But in a few years, who knows?"

Meanwhile, another type of behavior modification is needed for a large number of hypertensives who clearly need drug therapy but are not receiving it. As many as 40 to 50 percent of all who start drop out within a year. Since most hypertensives are free of symptoms most of the time, it is difficult, comments Dr. Campbell Moses, former President of the American Heart Association, to "make an asymptomatic patient feel better."

Three classes of drugs are used in treatment today. Although each can produce side effects — some trivial and bearable, others verging on the unacceptable — it is now possible to prescribe proper combinations containing lower dosages of each and thus minimize side reactions and maximize effectiveness. Indeed, so successful has drug therapy been for the moderate and severe hypertensive that, in the opinion of American Heart Association President Dr. Harriet Dustan, "the beneficial effects of treatment have diverted attention from basic research."

The cornerstones of therapy are the diuretics, which promote the excretion of excess salt and water. The development of the thiazide diuretics in the mid-1950s marked the arrival of the first specific anti-hypertensive medication and won the 1975 Lasker Award for its discoverers.[22] Many patients do extremely well on diuretics

[22] Drs. Karl H. Beyer, Jr., James M. Sprague, John E. Baer, and Frederick C. Novello of the Merck Sharpe and Dohme Research Laboratories.

alone, and even better when they limit their salt intake. Diuretics can be used in combination with other anti-hypertensive drugs to enhance their effectiveness.[23]

A second class of drugs — vasodilators — help reduce blood pressure by relaxing the smooth muscles of the blood-vessel walls, thus enlarging the channel through which the blood flows.

The third class of drugs works on the sites of ultimate blood-pressure control — the brain and other parts of the nervous system. Some work on the center in the lower brain from which orders go out to the heart to stimulate it to beat fast, pump more blood, and constrict blood vessels. This class includes drugs that appear to exert their action on the nerve cells that release the adrenalin-like chemical messengers (catecholamines) associated with stress and emotion.

One of the most exciting developments in recent years is the discovery of the beta-blockers, a group of drugs that act specifically to block the action of the catecholamines at the sites where they exert their effects. Originally developed to protect the heart muscle from excessive sympathetic nerve stimulation, they have been a literal lifesaver for many heart patients. Although it was known as far back as 1964 that they lower blood pressure as well, it was not until 1976 that propranolol was approved for hypertensive therapy in the United States.

Hailed as one of the greatest advances in the treatment

[23] Because diuretics also promote the excretion of potassium, which is necessary for good muscle function, care must be taken to maintain potassium levels, either through eating such foods as bananas and citrus fruit, or by prescription. Loss of potassium can be a special danger when diuretics are taken for weight reduction. Diuretics can also raise blood sugar and uric acid levels; if this should occur, your doctor will watch them to rule out diabetes and gout. New versions, some of which are not yet available in the United States, promise to eliminate some of the objectionable features.

of cardiovascular diseases since the eighteenth-century discovery of digitalis,[24] the beta-blockers are now being investigated for use against anxiety and migraine headaches; for cutting down withdrawal symptoms in opiate and amphetamine addicts; and to treat glaucoma.

As a research tool, the beta-blockers are helping to unravel the ongoing mystery of hypertension. One path leads to the kidney, which releases an enzyme — renin — into the bloodstream. Harmless at that stage, the renin is then converted through a series of steps into the most potent constrictor of blood vessels known — angiotensin II. It is now believed that propanolol works by reducing the amount of renin in the kidney. By following the renin-angiotensin axis, researchers have succeeded in developing drugs that disrupt the innocuous sequence before it becomes dangerous. One drug (saralasin) that is available now can be used only by intravenous injection and is limited to severe situations. However, it may not be too long before an effective oral drug will be available. The good news, pharmacologically, is the trend toward developing substances with molecules tailored to deal with selected mechanisms in the complex blood-pressure network.

The commitment made in 1972 by the National High Blood Pressure Education Program has maintained its momentum, with some changes in the light of five years' experience. Street-corner screening has given way to emphasis on detection of unidentified hypertensives in settings that provide confirmation and follow-up — doctors' offices, clinics, etc. "Awareness of high blood pressure," asserts NHLBI Director Dr. Robert Levy, "does nothing

[24] The 1976 Lasker Award went to Drs. Raymond P. Ahlquist and James W. Black for their discovery.

for an individual unless he or she acts on this information and achieves a normal blood pressure."

The year 1977 marked a milestone in American medicine — the first major consensus on an approach to the diagnosis and treatment of the disorder was developed, providing guidelines for identifying who needs treatment, when, what kind and how much. Backed by the U.S. Public Health Service, the Veterans' Administration, and seven major medical associations, the recommendations were embodied in the final report of the Joint National Committee on Detection, Evaluation and Treatment of High Blood Pressure, whose chairman, Dr. Marvin Moser, is senior consultant to the national program. Follow-up education within the profession is being vigorously pursued through a proliferation of postgraduate symposia for physicians, enlisting the talents of some of the leaders in hypertension from early childhood to old age.[25]

The commitment to the new priorities of genetic and childhood origin of hypertension is underscored by the awarding of sixteen NHLBI grants to run for two to five years. Under study will be blacks, whites, Spanish Americans and two Polynesian groups. Twins and family groups will be looked at for the role of genetic, environmental and psychological factors. One grant will correlate the relationship among blood pressure, dietary habits, lifestyle and characteristics of parents in Seventh Day Adventist elementary and high school students. To encourage the sharing of their findings and progress, the investigators will meet three times a year.

[25] Including Dr. Jennifer Loggie, pioneer in childhood hypertension; Dr. Ellin Lieberman, Pediatric Task Force member; Dr. Maria New; Dr. Edward Kass; Dr. Marvin Moser; Dr. Herbert Langford; and geneticist Dr. Henry Lynch.

Still another critical area is the nagging question of high blood pressure in blacks, among whom it is more serious and more widespread than in the rest of the population. The case for genetics is strong. In West Africa, where most American blacks originate, the rate of hypertension is the highest on that continent. A Louisiana State University team found significantly higher blood pressures among 10-year-old black children than among their non-black classmates. In some, the elevation was already discernible at the age of 5. Says a black physician: "Every woman in my family has had hypertension by the age of 45. Will I be next?"

In pursuit of a biochemical marker for high blood pressure, Dr. Kass and his associates are studying an enzyme in the urine — kallikrein. A powerful dilator of blood vessels, possibly one of nature's own tools to keep pressure down, it is low in the urine of adults with fixed hypertension. From urine specimens obtained from children, the researchers have found that those with the highest pressures have the lowest enzyme levels, while those with the lowest pressures are relatively higher in kallikrein. Those black children in the study whose pressures were not significantly different from the non-blacks' pressures tended to have lower levels. Why this is so is not yet clear, but it may turn out that, as in other genetic disorders, one clue is the enzyme deficiency.

On the other hand, the environmental stress to which blacks are subject is different from that of the rest of the population in both quality and quantity. The concept that "class counts more than race" is gaining growing acceptance. There is more hypertension among both blacks and whites in the lowest socio-economic groups. Among men who lost their jobs when their plant shut down, it was those who remained unemployed who suffered significant elevation in blood pressure — and the more severe the unemployment, the higher the pressure. There have been special stresses on blacks whose aspirations toward upward mobility have been blocked, with particular frustration experienced by the undereducated and undertrained.

Moreover, it is only in recent years that the gap in health-care delivery to all segments of the population has begun to narrow, with some of the most striking benefits achieved in the control of severe hypertension. "Blacks may not be genetically predisposed or vulnerable to malignant hypertension," says Cornell University Medical College's Dr. Michael H. Alderman. "They may simply be arriving too late for effective therapy, and then receiving inadequate treatment." At the Eighth International Epidemiological Association Meeting in the fall of 1977, Dr. Alderman reported that deaths from malignant hypertension had declined by almost 80 percent in New York City during the period 1958–74. "While the non-white rate was dramatically higher in the early years of the study, . . . it has declined four times as fast," he said, ". . . suggesting that in the future this racial difference may disappear."

Today's successes are by no means the final answer. To be sure, the well-treated hypertensive can realistically look forward to a future without a stroke, malignant hypertension or hypertensive heart disease. He can even, for the first time in history, qualify for life insurance. But high blood pressure remains an incurable disorder that can be contained only through lifelong adherence to not yet perfect medication.

Despite its flaws, today's therapy could have saved the lives of some of the country's most important political leaders of the twentieth century. President Woodrow Wilson suffered a severe stroke in 1919 — a crucial year for the success of his program to make the first world war the war to end all wars. Although he completed his term (to

1922), his capacity to carry on was sharply curtailed and he was virtually a president in name only. By contrast, President Franklin D. Roosevelt continued to lead the country with vigor and alertness out of a crippling depression and almost to the end of another world war, before his hypertension culminated in a massive cerebral hemorrhage at 1:15 P.M. and death at 3:30 P.M. on April 12, 1945.

The physician who saw him almost daily from March 1944 to the last moment of his life — and who accompanied him to Hawaii to chart the course of the war with Japan, and to Yalta to finish plans for the end of the war in Europe — gives an account of the President's health history,[26] starting in 1937, when the hypertension was first diagnosed. "The state of the President's health during the last year of his life was a topic of general concern," explains cardiologist Dr. Howard G. Bruenn, at the time a naval officer and consultant to the National Naval Medical Center. ". . . rumor and speculation were primarily its basis. This clinical record is written in the interest of accuracy and to answer some unfounded rumors."

The facts are that although the President's blood pressure fluctuated during the eight years of his illness from a high of 240/130 to a low of 170/88 (once), it was never at a normal level. However, he suffered no strokes, large or small; no impairment in memory, recent or past; and no speech difficulty. His first big complication was an attack of congestive heart failure in 1944, and the last was when he suddenly complained of a "terrific headache" on the morning of April 12, 1945.

"I have often wondered," Dr. Bruenn concludes, "what turn the subsequent course of history might have taken if the modern methods for the control of hypertension had been available."

[26] *Annals of Internal Medicine*, April 1970.

4

Diabetes — New Light on an Ancient Mystery

Last year will be remembered by Rosa, 52-year-old grandmother of three; Jane, 18-year-old college freshman; and Robbie, 15-year-old high school sophomore, as the time they all learned they are diabetics. Like the other four and a half million diabetics in the United States today who continue to raise families, pursue careers, and enjoy sports, all three are functioning well in their day-to-day lives. But, as Robbie describes it, "Life is more than bearable but less than beautiful."

Diabetes is not a glamour disease. A lifelong hereditary disorder of sugar metabolism, it rarely makes headlines, for it is associated with no dramatic triumphs and few dramatic defeats. Or so it seems to the general public.

The victims, their families and their doctors see it differently.

There is drama in the constant awareness with which the diabetic must live. "Granted, I don't think about it every minute of every waking hour," a 22-year-old college student writes to the Juvenile Diabetes Foundation as she looks back on her last ten years, "but it is on my mind many times a day." And while she is doing well now, "the worst part still exists — the fear of the future."

The fact is that with the added years of life granted to the diabetic since the momentous discovery of insulin in 1922, a large toll is exacted in damage to blood vessels, nerves, eyes, kidneys and heart. For many, diabetes still spells a shortened life span. Indeed, it is today the third-ranking cause of death, surpassed only by heart disease and cancer.

There is drama in the awesome dimensions of the problem, spelled out for the first time in the December 1975 report to Congress by the National Commission on Diabetes.[1] Up from 1½ million in 1950 to almost 5 million in 1975, it is now labeled a "major health problem affecting as many as 10 million Americans . . . and indirectly as many as 50 million who will pass the tendency to develop diabetes to their children, grandchildren, or both." Indeed, diabetes will be a threat to one in every five children born today, unless a cure or prevention is found in their lifetime.

Never has a new look at diabetes been needed so urgently as it is today. The euphoria that swept the medical world in the 1920s — when, thanks to insulin, the onset of severe diabetes no longer meant a death sentence in a year or two — has given way to frustration and disappointment. Old problems persist; new ones have appeared — and no major advance in treatment has been scored for more than half a century.

Today, with a sudden surge of new developments in

---

[1] Serving on the Commission are the Directors of seven of the National Institutes of Health; six nongovernment scientists and physicians, including some of the most distinguished leaders in the field; and four persons from the general public, at least two of whom are diabetics themselves or are parents of diabetics.

recent months and years, that long-awaited new look is here.

☐ Scientists are talking about a newly discovered brain chemical — somatostatin — the first new approach in treatment in more than fifty years, and called "twice blessed" by Harvard University's Dr. George F. Cahill, Jr., President of the American Diabetes Association.

☐ They are looking at the role of a hitherto obscure hormone from the pancreas — glucagon — now sharing centerstage with insulin.

☐ They are dealing with a completely revised view of juvenile diabetes, like Robbie's, that may well be the result of a viral infection, and will perhaps be preventable with a vaccine in the future.

☐ They are marveling at a new surgical procedure that restores sight to some diabetics after years of blindness, and the success of lasers in halting damage in the eyes of others threatened by blindness.

☐ They are following progress in spare-part organ transplants, and they are coming closer to the long-elusive genetic markers that will identify in the future who in Rosa's, Robbie's and Jane's families may be the next at risk.

Included in the 1970s look at diabetes is the concept that diabetes is not a single disease but a group of disorders sharing a common feature — high blood sugar. Each has its own pattern of inheritance, each with a different inherited susceptibility to a different trigger that sets it off, and each with its own impact on the life of the victim.

Although diabetes can strike at any age (it has been found in a nine-day-old baby boy and a 99-year-old woman), more than 80 percent of all diabetics are, like Rosa, over 45 when it surfaces (mature onset) and not the first in the family. Her sister and two cousins, all overweight since their early twenties (again like Rosa), are also diabetics. "I gained after each baby," she admits, "and never could take it off."

The first sign for Rosa was a persistent itch that embarrassed her during the day and disrupted her sleep at night. After her doctor confirms his suspicions,[2] he explains that hers is a mild disorder and that "with good sense, good will and a serious effort to get your weight down and keep it down," her life will not be too different.

To be sure, she must now live with a new awareness even of trivial events if she is to avoid complications. "Look at the shoes I am wearing to my niece's wedding tonight," she tells me. "A year ago I would not have worn them just to sit on the porch, but this morning I noticed an irritation on my foot and I am not taking any chances with new shoes and an infection." But the doctor is right — the biggest hardship she has suffered so far is in losing weight (more of that later); otherwise her life is by no means drastically changed.

Not so for Robbie, one of more than a million juvenile diabetics in the United States — an estimate that National Commission member and founding President of the Juvenile Diabetes Foundation, Lee Ducat, is certain is far too low. The number of juvenile diabetics is growing at an alarming rate. Juvenile diabetes was once considered a rarity (1–2500), but recent studies in various parts of the country reveal as much as a fourfold increase (1–600 in Michigan, 1–500 in Minnesota, 1–300 in the Philadelphia school system). One of the leading pediatricians in my town told me that after not seeing a single juvenile diabetic in twenty years of private practice, he discovered two within weeks of each other not too long ago. One is a

---

[2] With several blood and urine tests, including a glucose tolerance test that revealed how well she metabolized a measured amount of sugar.

10-year-old boy who suddenly began to wet his bed, and the other an 11-year-old suffering acute stomach pains.

Swelling the ranks of juvenile diabetes is a segment of the population in their forties, fifties and some in their sixties who have been the beneficiaries of insulin therapy from early childhood.

Robbie is the first known diabetic in his family, and there are *no* trivial events in his illness — it is far more serious than Rosa's. Home from camp with vacation time left before school started, he was putting in three or four hours a day on the tennis court, practicing for a county tournament. "Suddenly," his mother recalls, "he seemed to have an insatiable thirst, but it was a very warm September and with all his running around I wasn't too concerned."

Her complacency was abruptly and permanently shattered the morning he stumbled out of his room close to collapse, eyes glassy and a strange odor on his breath. She lost no time in getting him to the doctor, and the doctor lost no time in rushing him to the hospital. Robbie was in an impending coma which, uncontrolled, could have been fatal.

By the time he leaves the hospital, he and his family know that he faces a lifelong dependence on insulin, his diabetes the result of sudden and extensive damage to his insulin-producing cells in the pancreas. Like Rosa, he will continue to function well in his day-to-day life — he is making plans for college and is still playing tennis — but his is an illness that imposes a double burden, emotional as well as physiological.

Jane had no warning at all — her diabetes was picked up in a routine examination for her college application. Her disorder is mild, like Rosa's, and despite her age, she is not a juvenile diabetic. As far back as she can remember, her mother and grandmother have been mild diabetics. And as far back as they can remember, Jane has been a chubby child.

She is one of a new category of diabetics so recently described that there is, as yet, no estimate of how common it is, but because of its close similarity to mature-onset diabetes it is called MODY — Mature Onset Diabetes of Young People. Because of its similarity to mature-onset diabetes in its mildness, association with obesity, and familial tendency, researchers believe it may be a version of mature-onset diabetes that makes its presence known early in life.

The strong genetic aspect is highlighted in a 1975 study of MODYs by University of Michigan Drs. R. B. Tattersall and S. S. Fajans, who found that 85 percent had a diabetic parent, 53 percent of their brothers and sisters were MODYs, and 46 percent of the families were, like Jane, a three-generation phenomenon.

Nor does age alone determine juvenile diabetes. There are chronologically mature adults who are suddenly struck with the severe insulin-dependent disorder of the juvenile. And adding to the enigma of diabetes is the presence of abnormal sugar metabolism as a feature in about thirty relatively rare genetic diseases.

Familiarity with the "sweetness" disease dates back thousands of years. The Greeks gave it a name — diabetes — referring to the voluminous amounts of urine produced. By the eighteenth century doctors knew that when the excessive urine was excessively sweet, the patient was in trouble. In fact, a leading physician of the time reasoned that because so much sugar was lost in the urine, it would be a good idea to replenish it. He succeeded only in making a bad situation considerably worse.

The doctor had no way of knowing that because the diabetic is unable to use or store sugar, abnormal amounts accumulate in the blood, with the excess spilling over into the urine. He is literally starving internally in the midst of plenty.

The first clue as to why this happens emerged in 1889, when two experimental scientists, curious about the function of a long, narrow gland near the stomach, removed it. Without the pancreas, the animals developed the classical signs of diabetes. The search was soon on for the substance in the pancreas responsible. The hypothetical substance was even given a name — insulin.

Hypothesis became fact in 1922, when two young Canadian researchers, 30-year-old physician Dr. Frederick Banting and 22-year-old medical student Charles Best, announced that after only eight months of intensive work they had succeeded in preparing an extract from clusters of cells dispersed in the pancreas of animals (islets of Langerhans). When it was administered to dogs whose pancreas had been removed, the animals were once again free of symptoms of diabetes!

Results in the first patient, an extremely ill 11-year-old boy, were just as spectacular. Within months, insulin was made available for widespread use, winning instant acclaim by the medical profession and instant recognition by the Nobel Committee. Only one year was to pass between the discovery and the Nobel award, granted in 1923. (Ironically, the name of Dr. Charles Best was not included in the award.)

What is insulin's role? The sugars and starches you eat are broken down in your body to glucose which is then carried to each of your cells, where it is either used immediately for energy or stored in the liver as a reserve for future use. While glucose enters some cells readily — notably the brain — other cells (including liver, muscle and fat) need the aid of insulin for the glucose to penetrate. Insulin not only promotes entrance of glucose into the cells; it also stimulates the build-up of proteins from amino acids in the muscles and liver. Disruption of normal metabolism can show up in frequent infection (the basis for Rosa's itch), irritability, loss of weight, excessive thirst and excessive urination.

In uncontrolled diabetes, fats are broken down at an abnormally high rate with the release of by-products — ketones. If enough ketone bodies accumulate in the blood and spill over into the urine, the outcome may be coma — even death. This is keto-acidosis,[3] more of a threat to the severe juvenile diabetic like Robbie than to the milder mature-onset diabetic like Rosa or Jane. Indeed, that was Robbie's condition on that hot September morning.

The 1922 promise of insulin to prolong life has been abundantly fulfilled for the severe diabetic who no longer makes his own. Today, death from diabetic coma and keto-acidosis is an uncommon event. Over the years, however, an increasing suspicion developed that the overwhelming majority of diabetics — the mature-onset ones like Rosa — do *not* suffer an insulin deficiency. On the contrary, they often produce insulin in excessive amounts.[4]

Initial reaction to this disclosure was that perhaps the insulin is defective. Not so; it is competent insulin produced in adequate amounts. It was not until the 1970s that the answers began to come in. It is now known that the underlying defect is closely tied to an age-old characteristic of the mature-onset diabetic — obesity.

When the ancient Hindus described the "sweetness dis-

---

[3] Keto-acidosis has been precipitated in persons by a variety of fad reducing diets that feature excessive breakdown of fat tissue without suitable safeguards.

[4] This discovery by Drs. Samuel Berson and Rosalyn Yalow in 1960 and the technique that made it possible — radioimmunoassay — won the coveted Lasker Award in 1976 for Dr. Yalow, the first nuclear physicist and the first woman to be so honored. In 1977 Dr. Yalow was awarded a Nobel prize.

ease in the old and the fat and in the seed," they were close to the mark on the ongoing conspiracy between nature and nurture. The risk for mature-onset diabetes (more common in women than men, and in poor and non-white) doubles with every decade of life and every 20 percent of excess weight.

Just how much genes count is highlighted in a 1972 study of identical twins by Drs. Robert B. Tattersall and D. A. Pyke of London's Kings College Hospital. When one twin over the age of 45 comes down with the disease, there is almost a 100 percent probability of diabetes for the other twin. Indeed, if Rosa had an identical twin, it is virtually certain that within three years she too would be diabetic. But even with identical twins, environment counts, for when the first twin develops diabetes before the age of 45, the risk for the other twin is no more than 60 percent. (Almost half of the twins had a diabetic parent.)

Other studies reveal that when both parents are mature-onset diabetics, 60 percent of their offspring also suffer from a mild form of the disease by their sixtieth birthday.

With nature setting the stage, lifestyle and experience then take over. Given conditions of overnutrition and underactivity, incidence of diabetes goes up. So it is with the Pima Indians of Arizona, who combine an inherited ethnic vulnerability with a sedentary lifestyle on the reservation plus adoption of a typical "American" diet. Close to 50 percent of the men and women over 30 have abnormal sugar metabolism, and the rate at which they suffer from eye and kidney complications associated with diabetes is ten to fifteen times higher than that of the rest of the population.

In Europe, diabetes has declined with the food shortages of every war in the last century, only to rise again with peace and relative affluence. The most striking rise has been in the most affluent countries, notably the United States and West Germany!

You are at a particularly high risk if you are genetically vulnerable, obese in middle age, and were markedly overweight as a teenager. A 1970 study of 85,000 obese women conducted by scientists at the Medical College of Wisconsin with TOPS (Take Off Pounds Sensibly) revealed that those who were obese teenagers were twice as likely to develop mature-onset diabetes as were women who were less than 5 percent overweight as teenagers. The majority, however, like Rosa, gained their extra pounds between the ages of 25 and 44.

The link between obesity and the failure to use insulin began to emerge with new insights in basic research that came to light in the early 1970s. For the insulin to bind to the cell where it is needed, it must fit into a specific receptor, much as a key fits into a lock.[5] The pioneering work of NIH researchers Drs. Jesse Roth and C. Ronald Kahn and Stanford University's Drs. Jerrold M. Olefsky and Gerald M. Reaven disclosed that the problem in the obese mature-onset diabetic is too few receptors and, in some instances, diminished capability of the receptors to recognize the hormone.

Normally, your blood sugar rises after you ingest and metabolize carbohydrates. Your body responds by increasing the output of insulin (a small level is always present) to help the glucose enter the cell. In the obese, both mouse and man, the insulin is powerless because there are too few receptors — 50 percent fewer, in fact.

---

[5] Studies of receptors and the role they play in crucial life processes are in the frontiers of science today — so new that the *New England Journal of Medicine* in April 1976 remarked about the "older literature" on the subject (three or four years ago) and then commented "older indeed"!

With the sugar level still high in the blood, the pancreas pours out more insulin. Still to no avail. It is now a self-defeating and possibly harmful cycle, for the continued futile response may lead to exhaustion of the pancreas.

The next chapter in the receptor story is one with enormous implications for treatment of the obese diabetic. By cutting down on calories — losing weight and keeping it down — the process can be reversed! The number of receptors increases, insulin is produced in normal amounts, and glucose can now penetrate the cells.

Long before receptors, long before insulin, doctors already knew that weight reduction alone could control obese diabetics. In 1796, Scottish physician Dr. John Rollo treated British Army Captain Meredith, "a corpulent man and a diabetic," with a diet of "rancid meat" and meager calories. Captain Meredith lost his symptoms, his "saccharine urine," his corpulence, and soon returned to active duty. The findings were duly published in 1798 in *A General View of the History and Appropriate Treatment of the Diabetes Mellitus.*

Unfortunately, not all diabetics today share Captain Meredith's unswerving adherence to a diet, nor do all doctors have Dr. Rollo's persistence and faith in pushing it.

The problems with losing weight and keeping it off with diet alone are familiar to millions of Americans for whom obesity is a sinister partner not only in diabetes but in heart disease and high blood pressure as well. And so, in the 1950s, when pills were developed capable of bringing blood sugar down in mature diabetics who did not need insulin, they were eagerly welcomed. (One type is designed for the obese, another for the lean.) Today, Rosa is among the 1½ million Americans on the anti-diabetic pill.

In the early 1970s, the oral drugs came under a cloud of suspicion when a study by the University Group Diabetes

Program questioned both their safety and their efficacy after prolonged use. While some of the leading diabetes experts in the country believe these conclusions unwarranted, others are looking with renewed interest to the 200-year-old cure — diet and loss of weight.

Can it be made to work in the face of repeated failures? The experience of Emory University's Dr. John K. Davidson, at Atlanta's Grady Memorial Hospital, is a resounding YES. Since 1971, oral drugs have been completely abandoned, and for those who still need insulin, the amount has been significantly decreased. All the obese, including one woman who had to lose 100 pounds, are down to an acceptable weight, with their sugar in good control. All are generally healthier than ever.

The key to success, Dr. Davidson maintains, is "intensive instruction and continuous follow-up." It begins with a full day at the Diabetes Day Care Center for the patient and often a member of the family. A team, including the physician, teaching nurse, dietitian, podiatrist and others, work with him and his special problem. Where indicated, consultations are arranged with other specialists — eye, kidney, vascular, etc.

Although calorie restriction is stringent (including, for some, a one-week fast under careful supervision), the diet reflects the individual preference of the patient. Especially helpful both to the patient and the family is easy access to the professional team when a problem arises.

In October 1976, Dr. Davidson helped spread the message to 400 colleagues from all over the country, when he was enthusiastically received at Diabetes Day, a new venture in professional education by the NIH, designed to speed the transfer of new and important information from the research laboratory to the bedside.

Rosa's doctor is among those who are willing to give Dr. Davidson's venture a chance, if Rosa agrees too. Mindful of the too frequent occurrence of diabetes in her family, she is also concerned about the future for her children: her 28-year-old daughter, size 16, who dreams of fitting into a 12; her 26-year-old son, who is no longer the athlete he was in high school, but is still eating as if he were; and her youngest daughter, still slim at 22, and determined to remain that way, not for health reasons but because she knows that a fat girl has a poor future in the world of fashion design.

A 1976 discovery by Georgetown University Professor of Biochemistry Dr. Melvin Blecher and his colleague Dr. Stephen Goldstein promises to identify who among them is at risk. The search to identify the potential or pre-diabetic has been on for many years. Several investigators report early damage to blood vessels that can be seen with the electron microscope. Another test is a delayed response to insulin. Neither is a consistent predictor, however.

While studying receptor activity in mature-onset diabetes, the Georgetown researchers confirmed that not only is the capacity to bind insulin diminished, but the capacity to bind glucagon (the other hormone of diabetes) is also reduced. When they looked beyond the diabetics to their children, they found that nine out of ten, *with no sign or symptom of the disease*, had fewer receptors, a condition hitherto expected only in an overt diabetic. Indeed, the level of hormone they could bind was only one-third that of normal individuals with no family history of diabetes.

Dr. Blecher says he has been inundated with inquiries from doctors since the publication of his findings, but routine clinical application will be feasible only when the still very complex test has been simplified and its validity established in larger numbers of families.

For Rosa and her diabetic relatives, their genes need clearly *not* be their destiny. The problem is not that simple for the insulin-dependent, especially the young. Karen, at 16, a five-year veteran of diabetes, is packing for a ski weekend and she needs no reminder of what goes into her valise: insulin to keep her blood sugar down; glucagon, a substance that will raise her blood sugar if insulin brings it down too much; syringe; etc.

"Will I be blind when I'm 30?" she suddenly asks her mother.

Like many juvenile diabetics, Karen is not only extremely well-disciplined in self-care, she is also extremely knowledgeable about the disease and its possible complications. She knows that long-term diabetes is the fastest growing cause of new blindness in adults in the United States today,[6] and while the odds that Karen will *not* be blind at 30 or 40 or 50 are strongly in her favor, they are not as good as for the 16-year-old who is *not* a diabetic.

Once a rarity, diabetic retinopathy (damage to the light-sensitive layer of the eye) is now occurring with increasing frequency, paralleling the increased life span of the diabetic. Nevertheless, it is by no means inevitable. Reassurance comes from Johns Hopkins' Dr. Stuart L. Fine, who explains in the spring 1976 *Sight Saving Review*, "not all diabetic patients develop retinopathy and . . . it is *important to remember* that only a *fraction* with *retinopathy ever develop* serious problems including blindness."

While she can cope with the increased risk of heart disease, kidney failure, nerve damage, she views the threat to vision as the most ominous, dangers not exclusive to the juvenile diabetic. A 1976 report in the *British Medical Journal* warns that one in every five mature-onset diabetics already has early recognizable complications by the time he first seeks medical advice, but the brunt is borne by the juvenile, and long years of severe impairment.

[6] There are today 48,000 legally blind from diabetic retinopathy and more than 300,000 threatened with blindness.

Karen's concern about the threat to vision is shared by many of the long-term diabetics with whom I spoke, but other nagging questions also come up again and again. From both young men and young women: "How about raising a family?" From the wife of a 42-year-old diabetic: "He is beginning to worry about impotence." From a 26-year-old who says that despite "careful juggling of insulin, diet and exercise I seem to drop the ball too often." They ask about "what pot will do to my blood sugar," and demand no-nonsense answers about the promised new treatments ahead. A recurrent query, especially from the young: "Why me?"

Nineteen seventy-six was a landmark year for the diabetic whose greatest concern for the future is his vision. In an unprecedented information program designed to reach the general public through a special press briefing and the medical profession through a special mailing, Dr. Karl Kupfer, Director of the National Eye Institute, revealed that it is now possible to reduce the risk of blindness due to retinopathy by almost 60 percent. The tools are powerful beams of light — green argon laser and white xenon arc. The technique is photocoagulation.

Indeed, so striking were the results in the first two years of a ten-year study that advance copies of the report were sent out to 10,000 ophthalmologists and more than 300 physician members of the American Diabetes Association.

In April 1978, the NEI announced a new study involving twenty-two centers that will seek to determine whether treatment at an earlier stage of retinopathy may be of value in reducing the risk of blindness, and if so whether this benefit outweighs the risk of adverse side effects. Investigators also hope to find out if the laser treatment is effective against macular edema (an abnormal accumulation of fluid in the retina), which often accompanies diabetic retinopathy and may cause blurred central vision. A third objective is to find out if aspirin, alone or in combination with another drug, may be useful in treating diabetic retinopathy. Because diabetics have an increased tendency for platelets to clump, the small blood vessels in the retina may be more likely to become blocked. Since aspirin is known to interfere with platelet clumping, the rationale in this program is similar to that in the aspirin study in CHD. (See page 77, Chapter 2.)

Retinopathy is an umbrella term describing abnormal changes in the blood vessels of the retina. Although most commonly found in diabetics of long standing, it can also occur as a complication in persons with high blood pressure, pernicious anemia and other disorders. The most common version involves changes in the tiny blood vessels — microaneurisms — confined within the retina. Hundreds of thousands of diabetics can endure this condition for many years without suffering any serious impairment of vision. When, however, it progresses to the extent that large areas of the retina are deprived of normal blood supply, new blood vessels develop to replenish the supply.

If the new vessels begin to grow on the surface of the retina and into the clear gel-like vitreous (proliferative retinopathy), a new danger develops — hemorrhage into the vitreous and serious detachment of the retina now becomes a permanent threat to sight. This is the condition that photocoagulation is designed to correct.

When the NEI study was initiated in 1971 at sixteen medical centers across the country, with 1,700 patients participating, the photocoagulation treatment was limited to only one eye. The two-year follow-up revealed that (1) the majority of the eyes, both treated and untreated, did *not* go blind; and (2) the percentage that did go blind was significantly greater among the untreated (16.3 percent) than among the treated (6.4 percent). So the decision now is to reexamine each patient to see if the untreated eye can also benefit from photocoagulation. The plea to the medical profession is: diagnose and identify the patient early who is a candidate for the procedure.[7]

Within weeks, a second study was unveiled. Can vision be restored to persons already blind because of recurrent bleeding into the vitreous that has failed to clear up by itself?

With the development in 1972 of new instrumentation by University of Miami's Dr. Robert Machemer, the hitherto unapproachable vitreous is no longer off-limits to the specially trained surgeon. It is now possible to remove the cloudy, blood-filled vitreous and replace it with a clear solution.

Because vitrectomy is not without its own hazards, it has been the practice to wait for at least a year after the hemorrhage, to give the eye as long as possible to heal itself. Now, however, experience suggests that it is better not to wait. The NEI is now seeking to enroll between 600 and 700 patients in thirteen medical centers, with the requirement for admission being a severe vitreous hemorrhage within the preceding five months.

One of the most stunning successes of vitrectomy is that of a long-term diabetic in her forties, a mother of two who had been blind for seven years. Perhaps the most memorable moment of her life was when she suddenly saw her boys as adolescents, after having "seen" them grow into those crucial years only through their voices. Today, her vision is almost perfect — 20/25 in the first eye within six months after surgery, and 20/40 in the second eye. In fact, she is driving again.

---

[7] A similar message based on British studies was directed to the medical profession in Britain, with a plea in the March 19, 1977 *British Medical Journal* to ". . . emphasize with new urgency the need for early diagnosis and evaluation . . ." so that those who can profit from photocoagulation will receive it in time to avert permanent damage.

For a 28-year-old artist whom I met recently — a diabetic since fifteen — the final results are not yet known. "For years I've known everything I could expect," she tells me, "but I never thought my eyes would go so soon." And when it happened it was sudden and complete — "like a thick curtain of vaseline pulled over my eyes." She was scheduled for vitrectomy within two weeks after the blindness struck, and she glows as she describes what it means when "the eye suddenly opened up three days after surgery." She sees!

Hers is a particularly severe problem — photocoagulation on both eyes was unsuccessful — but the doctors are now optimistic about a good result with the eye where such fast action was taken. Not ready to read yet or to pick up her paintbrushes, her vision is good enough for her to serve us lunch and fill me in on the stories surrounding the many interesting artifacts that she and her husband have collected and displayed with rare good taste.

In an interview with Harvard University's Dr. Felipe Tolentino, of the Eye Research Institute, Retina Foundation, Boston, he explains who can profit from this bold new procedure. "First we must make sure that there is still light perception, that there is not too much damage to the retina." And to determine if the retina is in good shape there are sophisticated aids, including ultrasonics and electroretinograms which can pick up a response to a flash of a bright light through electrodes on the blind eye.

In 1975, Dr. Tolentino added an extra dimension to the efficacy of this challenging procedure when he published the description of an improved instrument developed at the Eye Institute. Not only is it automated to perform better, as Dr. Tolentino points out, but because it "gives the surgeon complete control of the instrument, he need no longer relay instructions to an assistant, giving the surgeon the advantage of time."

It has been known for many years that diabetics are

particularly vulnerable to cataracts (opacity of the lens),[8] often at an earlier age than non-diabetics. Although practically no one need suffer lack of vision due to cataracts, for they can be removed surgically, safely and effectively, it would be far better to prevent them or slow down their development.

Different types of cataracts, scientists now know, develop by way of different mechanisms. Today, the first success in delaying development of a cataract has been achieved with the "sugar" cataract of a diabetic; the first beneficiary — a degu, a tiny rodent from high up in the Andes.

The excessive sugar in the lens of the diabetic is converted by the enzyme aldose reductase to the sugar alcohol sorbitol, which in turn triggers the development of the cataract. Dr. Jin H. Kinoshita and his associates at the National Eye Institute's Laboratory of Vision Research reasoned that if the enzyme could be inhibited, the sorbitol would not accumulate. In their search for the appropriate agent, they hit upon a plant extract — a flavonoid. It was successful in preventing formation of cataracts in animal lenses in test-tube studies, and in 1974 they reported delayed formation of cataracts in the eyes of living animals with "sugar" cataracts caused by a rare genetic disease — galactosemia.

The scientists needed a better animal model — one with diabetes and the tendency to develop cataracts at the same stage of life as the human diabetic does. The degu met these requirements.

In 1977 the research team reported that within ten days after the degu becomes diabetic, it develops cataracts. When, however, the tiny animal is treated with the powerful enzyme inhibitor quercitrin, he is still free of cataracts twenty-five days later, despite a blood sugar

[8] See page 186, Chapter 6.

level as high as that of his untreated counterpart. While it is still too early to conclude that quercitrin can *prevent* a "sugar" cataract, there is no question it has the power to delay it.

"Sugar" cataracts account for only a relatively small proportion of cataract victims in the United States today. And what works for the degu may not do as well for man. Meanwhile, these findings have spurred continued investigation into the role of other enzymes in cataract formation, particularly in a strain of non-diabetic mice whose heredity makes them prone to cataracts early in life. As the research progresses, scientists are hopeful that new clues will emerge to help the most common of cataract victims — the 3 million aging and not-so-old with "senile" cataracts.

Another reassuring weapon against "fear of the future" comes from new insights in the fight against heart attacks, to which the diabetic is at a twofold risk. When Jackie Robinson died in 1972 at the age of 53, the headlines said "heart attack" (his third in four years), but the doctors who cared for him knew that underlying the heart attacks were twenty years of diabetes.

Like Robinson, diabetics are likely to suffer heart attacks earlier in life, have high blood pressure more frequently, and go downhill more rapidly. Indeed, the advantage that pre-menopausal women enjoy in staving off heart attacks is lost to the diabetic woman.

Contributing to the risk for early heart disease is an increased cholesterol in a particularly dangerous form — low-density lipoprotein (see Chapter 2). In a pioneering study[9] as far back as 1955, one of us (I.J.G.) and my colleagues at the then Beth-El Hospital, Brooklyn, New

[9] Part of an award-winning exhibit at the 1955 American Medical Association Annual Meeting.

York, found that among juvenile diabetics ages 9 to 15, almost half already showed a significant increase.

Today, doctors know that the risks of high cholesterol and high blood pressure (see Chapter 3) can be diminished with diet and drugs and, as in the total program in prevention of heart disease, early identification is a must.

The kidney is a particularly vulnerable target in a diabetic, with a rate of disease 17 times as great as in the general population. Many for whom the damage would have been fatal in the past are alive and functioning today, some with the help of dialysis, others (more than 200) with kidney transplants. A report in the *Journal of the American Medical Association*, March 14, 1977, on 40 patients who received kidneys between 1970–75 at the Mayo Clinic, describes 19 as "fully rehabilitated."

With fewer kidneys available than patients who need them, the Mayo doctors comment that in choosing a candidate "a strong desire to live was a most important factor in decision-making," a quality I was to encounter in almost every diabetic with whom I spoke.

There is new hope today that a full and satisfying sex life can be restored to the large number of diabetic men who experience problems with erection. For some, it is impotence associated with neuropathy (damage to nerve cells). For others, as in non-diabetic men, the disorder is psychological.

Until recently, both doctors and patients were skeptical that the psychological factor is important in the diabetic. It was too easy to attribute the problem to organic damage. Sleep studies today, however, confirm its role. Two New York researchers, Mt. Sinai's Dr. Harold Rifkin and Montefiore's Dr. Howard Rothwag, both report normal erections in diabetics during sleep. Another investigator even found it helpful to wake the doubting subject at that moment to convince him that it was indeed happening.

For the middle-aged man whose impotence is caused

by neuropathy, there is help from the growing use of a surgically implanted prosthesis in the penis.

Basic research now in progress may provide a means of blunting neuropathy. The same sequence of events that triggers "sugar" cataracts (increased blood sugar converted to the sugar alcohol sorbitol by the enzyme aldose reductase) also takes place in other tissues and may contribute to neuropathy. A new drug — alrestatin — that promises to inhibit the culprit enzyme is now under study.

With so many organs and systems affected in diabetes, doctors have long wondered if the same genes that disrupt sugar metabolism also dictate whether eyes, kidneys, blood vessels, nerves, etc., will be particularly vulnerable to damage. Or are the complications the outcome of years of abnormal sugar levels producing an internal environment hostile and harmful to the target organs?

A 1976 report in *Diabetes* describes the experience of a series of patients who have received kidney transplants. Over a period of years, the diabetic recipients began to show changes in the transplanted kidney that were more intense and more frequent than those of the non-diabetic recipients. And when kidneys from normal rats are transplanted into diabetic rats, the healthy kidneys ultimately become severely damaged. The good news is that *when a diabetic kidney is transplanted into a normal animal, the diseased organ*, now in a normal internal environment, *begins to return to normal*.

At Boston's Joslin Clinic, Dr. Priscilla White followed the progress of 73 juvenile diabetics who developed the disorder before the age of 15 and lived with it for at least forty years. By traditional thinking, all should have suffered major complications. Not so. In a report to the 1974 American Diabetes Association meeting, Dr. Aldo T. Paz-Guevara and Dr. White disclosed that of the 73, 64 still had good vision, 58 had good kidneys, and 62 were free of coronary heart attacks. Furthermore, after forty years of diabetes, only 3 were unable to work because of the disease. For decades, each had followed a rigid regimen aimed at achieving blood sugar levels as close to normal as possible.

In the spring of 1976, the American Diabetes Association took an unprecedented stand on the long-controversial issue of maintaining good blood sugar control.[10] If complications to the eyes and kidneys in the young and middle-aged (those at greatest risk) are to be staved off, blood sugar levels close to the normal range are a must, according to Dr. George F. Cahill, Jr., President of the American Diabetes Association; University of Minnesota's Dr. Donnell D. Etzwiller; and Northwestern University's Dr. Norbert Freinkel, in an official release to the medical profession.

Until very recently, the gap between adopting the policy and implementing it would have seemed too large to bridge. First because, with the current state of therapy, optimum control still remains elusive for many, and secondly because it has not been possible to get a good picture of the extent of high blood sugar in a patient over a long period of time.

Now, thanks to an ingenious test developed at Boston's Children's Hospital Medical Center, one blood sample can tell the doctor what the average blood sugar level was in the preceding few weeks. Repeated measurements will then provide an objective estimate of a diabetic's blood sugar content over a period of years.[11]

[10] The decision was based on studies in the last five years on mice, monkeys, rats and Chinese hamsters, showing that prolonged periods of excessively high blood sugar levels in blood vessels that nurture the eye and the kidney spell trouble.

[11] As each new red blood cell enters the circulation for its 120-day

Harvard Professor Dr. Kenneth H. Gabbay, of the endocrine division of Children's Hospital Medical Center, talked about the new test's significance at the 1977 American Heart Association Fourth Science Writers' Forum in San Antonio, Texas. "In studies directly related to heart disease, we found that patients with poor sugar control have significantly increased levels of cholesterol" — the first direct link between the diabetic's metabolic disorder and the familiar risk factor for heart disease.

"We are now embarked on a five-year study of 100 diabetics to find out if poor sugar control is related to small blood vessel disease [accounting for damage to eye and kidney]." Indeed, this study promises to yield important new insights into the development of all complications as they relate to blood sugar level.

The test is now performed routinely in Dr. Gabbay's laboratory and is in use on a research basis in a dozen medical centers across the country. Not the least of its value is in evaluating the efficiency of insulin and oral drugs, as well as other new therapies as they develop.

Another mystery now being unraveled with the new look at diabetes is why good sugar control remains elusive for many — even for those who follow all instructions and break no rules. Robbie's mother, who has two older sons, expected some rebellion at his age, but she asserts that "he knows full well never to mess around where his insulin is concerned."

Over the years, there was a growing suspicion that more than lack of insulin is involved in the disorder. In the early 1970s, another hormone in the pancreas, glucagon,

journey, some of the glucose in the blood attaches to the hemoglobin, where it remains for the lifetime of the cell. Normally, about 5 to 7 percent of the hemoglobin is so altered. In the diabetic, it may be 10 to 20 percent, depending on the blood sugar level at the time.

emerged from the obscurity it had suffered for decades to share the spotlight with insulin. To Dr. Roger H. Unger and his co-workers at the University of Texas Southwestern Medical School, Dallas, goes the credit for demonstrating that insulin is not the sole problem. Glucagon is a substantial contributor.

While insulin acts to keep blood sugar down, glucagon does exactly the opposite — it raises blood sugar levels. Moreover, Dr. Unger found that in every diabetic he studied, both human and animal, high blood sugar level was accompanied by a high level of glucagon. Even more provocative was the evidence that it is not the high blood sugar alone that brings on the keto-acidosis; it is the increased glucagon. The higher the level of glucagon, the more out of control the sugar level.[12]

Interesting as Dr. Unger's observations were, their usefulness in helping the diabetic depended on finding a way to control glucagon. As so often happens in science, a chance discovery in a laboratory not on the trail of diabetes at the time provided the answer.

Dr. Roger C. L. Guillemin and his team of physiologists and biochemists at the Salk Institute, San Diego, California, had already gained worldwide fame for their work with brain hormones. By 1971 they had isolated, identified and synthesized, in the laboratory, substances (more than a quarter of a million sheep brains were the source) capable of controlling thyroid activity, milk production and fertility. The researchers then began to look for a substance in the brain that could stimulate growth hormone, urgently needed for the treatment of dwarfs.

[12] Because glucagon rises during starvation and with a high protein meal, it can be a hazard to the obese potential diabetic on some of the fad reducing diets where a high glucagon can unleash keto-acidosis.

Before a year had passed, they did indeed find a hitherto unknown chemical in the brain, but when they tested it, much to their disappointment, it did not increase growth-hormone production. On the contrary, it depressed growth-hormone production. By this time, it too had been purified, identified and synthesized in the laboratory, and it was named somatostatin.

In 1977, Dr. Guillemin shared the Nobel prize with Dr. Andrew Schally, Veterans Hospital, New Orleans, crowning a twenty-one-year race between the two scientists in pursuit of brain hormones.

The first clue that somatostatin affects sugar metabolism came in 1973 from the Primate Center in Seattle, Washington, when Drs. Charles Gale and John W. Ensick gave it to baboons and found that it lowered the animals' blood sugar. Their initial explanation was that somatostatin increases insulin. Not so. Actually, the insulin is decreased. But what was really a surprise was that glucagon *is decreased even more.*

Dr. Unger in Dallas then had the opportunity to find out what happens to diabetic animals whose glucagon is suppressed. The answer is history. Diabetic dogs from whom insulin was withheld — and who by all the old concepts should have continued to suffer dangerously high blood sugars — were brought back to normal with somatostatin, the first new substance in fifty-two years with that power.

The first test with human volunteers was in early 1974, when Dr. John E. Gerich and his colleagues at the University of California Hospital, San Francisco, gave somatostatin to 10 diabetics — the youngest was 21 and the oldest 56 — all dependent on insulin. Not only was the glucagon brought down, but the blood sugar went down with it. The effect was transient to be sure, but this was not entirely unexpected, since the early samples of somatostatin had short-lived activity. By 1976, a version that acts for as long as eight hours was available, in contrast to the first product, whose activity lasted no more than one hour. Particularly striking was the success of a combination of somatostatin and insulin in keeping in check the excessively high blood sugar level that often threatens the severe diabetic after a meal.

Dr. Gerich was soon to confirm in man what Dr. Unger had observed in animals. "Insulin lack per se," he reported in the *New England Journal of Medicine* in May 1975, "does not lead to fulminating diabetic keto-acidosis . . . glucagon is a prerequisite."

The greatest promise of somatostatin is for the young, who both are the hardest to control and suffer greatest risk of complications in the future. For them it can be what Dr. Cahill calls "twice blessed." Because growth hormone has been implicated in development of damage to the eyes in diabetics of long standing, somatostatin's power to inhibit growth hormone may also halt eye damage.

"More than 300 insulin-dependent patients have received the drug in trials," Dr. Guillemin told me at a November 1976 conference sponsored by the New York Academy of Sciences, "and at least one person has been kept under control with somatostatin alone for three weeks." He stressed the absence of harmful side effects during the periods of the trial so far. "Before it can be put into general use," he pointed out, "we must know more about what happens in long-term use — not only in control of blood sugar but in the rest of the body as well."

Somatostatin has so captured the interest of the scientific and medical world that in the brief time since Dr. Paul Brazeau (now at McGill University) and his colleagues at Salk reported their discovery in *Science* in 1973, more than 300 papers about it have been published. "It is setting the endocrine world on its ear," a West Coast researcher told me.

We now know that somatostatin is not limited to the

brain. It is also found in other parts of the nervous system, as well as in the pancreas and the intestinal tract, and its importance goes beyond its role in diabetes. It plays a role in stress, influences thyroid function, can prevent blood platelets from aggregating (a factor in heart disease), and brings about mood changes when given to rats.

With the 1977 landmark developments in recombinant DNA,[13] and the discovery that a harmless bacterium can be turned into a factory for producing somatostatin, hitherto undreamed of quantities may soon be available for research in diabetes and other endocrine disorders. The five milligrams of the hormone that Dr. Guillemin isolated from a half million sheep brains can now be obtained from two gallons of bacterial culture.

The increased availability of nature's form of somatostatin made possible by this dramatic new development will not, however, provide a product for widespread clinical use now. In a recent (August 1978) interview with Dr. Wylie Vale of the Salk Institute, who is one of the leaders of the scientist team that originally identified and synthesized the new hormone, he told us that "what we need now is a version that has more specific and more sustained effects. Our experimental progress in that direction is very promising."

For his role in opening up this new look at diabetes, in 1975 Dr. Unger received the prestigious Banting Medal, the highest scientific honor bestowed by the American Diabetes Association and one of the highest honors in metabolic disease research. The discovery, Dr. Unger said, "opens up a whole new avenue for investigation and control of diabetes by an entirely new approach. It really does raise tremendous hope for a new form of therapy."

[13] See page 12, Chapter 1.

Meanwhile, a decade of research aimed at the "ideal" solution — a spare-part organ — is beginning to yield results. By 1975, of the 46 transplants of healthy pancreas to severe diabetics, four had been functioning ten months or more, releasing insulin when needed and maintaining acceptable blood sugar levels. The procedure, however, has all the problems of other organ transplants — high rate of rejection and low supply of suitable donor organs. Furthermore, there is no evidence that it can halt complications that accompany long-term, severe diabetes.

A more promising approach is transplantation of only those cells in the pancreas that actually produce insulin (beta cells or islets). After ten years of worldwide research with animals, a team of University of Minnesota scientists reported in 1976 the first successful transplants in four men and three women, ages 23 to 49. "Insulin requirements for these patients," Dr. John S. Najarian told the Sixth International Congress of the Transplantation Society, "were reduced by 50 percent for nearly eighteen months."

There is still a way to go before this dramatic therapy will be practical. Islet cells are easily rejected — Dr. Najarian's patients are a special exception. They are all extremely ill, their diabetes so advanced that they have already received kidney transplants, and, like all patients who are struggling to retain a transplanted organ, are receiving medication designed to suppress their immune system. Hence, they are not as prone to reject the new transplant.

Another problem is how to obtain enough islet cells, for they are widely dispersed in the pancreas and account for only 2 percent of that organ. Improved techniques for storing and maintaining the cells until ready for use and for injecting them where they will provide the greatest benefit are all under study now.

Acknowledging that none of his patients were cured by the transplant, Dr. Najarian emphasizes that now we know "it can be made to work."

Still another approach is a computerized artificial pancreas, stocked with insulin and capable of delivering the hormone automatically to the diabetic at the moment he or she needs it. The first successful artificial pancreas was developed in Canada and West Germany. It uses electrodes to record continuous blood sugar levels, feeds that information into a computer attached to a pump, and then injects the appropriate amount of insulin.

This device, however, is the size of a TV set and can be used only at the bedside. What is needed now is a miniaturized version, one small enough to be implanted and to work for the patient as he goes about his routine activities. In the spring of 1977, scientists at the Medical Technology Division of A. G. Siemens, in Germany, demonstrated the first component of an "artificial beta cell",[14] a device not much larger than a matchbox, containing a miniaturized pump, a supply of insulin, a power pack and electronic parts. Another small box selects the rate at which insulin will be infused. When perfected, the apparatus should be small enough to be implanted, capable of storing enough insulin for 250 days, and accurate enough to deliver the required amounts on demand.

Despite the many problems still to be solved, a spare-part organ may not be too far off. Early in 1977, *Medical World News* revealed the responses to a poll conducted among 23 clinical authorities in a variety of specialties. When each was asked "What two or three useful clinical advances in *your* specialty do you anticipate will become conventional medical practice, and by what year?" the diabetes specialist, Boston's Dr. Priscilla White,

[14] Reported in *Medical Tribune*, April 13, 1977.

prophesied, "living beta cell transplantation and artificial beta cells," and both by 1980!

What are the prospects for restoring the capacity of the damaged pancreas to make its own insulin so that the diabetic will no longer need animal insulin[15] or spare-part organs, biological or mechanical? Research now in progress by the University of Chicago's Dr. Donald F. Steiner, 1976 Banting Award winner, may make it possible to decode the messages in the DNA of the genes that dictate the production of insulin and its two precursors, proinsulin and preproinsulin (recently discovered in Dr. Steiner's laboratory). Dr. Steiner foresees the day when such knowledge can be used to correct the basic genetic defect in the diabetic — a feat of genetic engineering at its highest potential.

Until not too long ago, the juvenile diabetic who asked "Why me?" would have found the doctor just as baffled. The sudden and unexplained onset resembled an infection more than a metabolic disorder, but no one had been able to pin it down. Indeed, as far back as 1864, there were scattered accounts of juvenile diabetes following mumps.

Over the years, circumstances bolstering the infection theory continued to crop up, but the first solid clue linking a particular virus to juvenile diabetes came from Britain, where Drs. D. R. Gamble and K. W. Taylor reported in 1969 that they found evidence of recent infection with a common flu-like virus, Coxsackie B4, in young people

[15] Of which we have a diminishing supply in the face of a constantly increasing demand. In the future it may no longer be necessary to depend on pancreatic extracts from animals for insulin. With the recent success in splicing a synthetic gene with instructions to make human insulin into a bacterium, the hope is that the *E. coli* will become a factory for the production of the desperately needed hormone (see page 3, Chapter 1).

who had recently developed diabetes. At about the same time they also noted a seasonal increase in new cases of diabetes that reflected a seasonal outbreak of the same Coxsackie virus.

Since then, other viruses and clusters of new cases have been reported in other parts of the world. In the wake of the 1964 worldwide epidemic of German measles, Australian doctors discovered that 20 percent of the babies born with the infection went on to develop either overt diabetes or abnormal glucose tolerance.

Playing the role of medical detective, as epidemiologists often do, Dr. Harry A. Stultz, Professor at State University of New York at Buffalo, delved into the records of Erie County Hospital and private pediatricians from 1945 to 1971. During this twenty-five-year period there were three large outbreaks of mumps, one every seven years. Paralleling these highs but occurring about four years later were peak numbers of new cases of juvenile diabetes!

When he interviewed the parents of a large number of the diabetic children, Dr. Stultz learned that 50 percent had been exposed to or been affected by mumps about four years before the diabetes appeared. He noticed a particularly sharp rise in new juvenile diabetes in the decade 1950–60, which he thinks might be attributed to the then popular practice of exposing preadolescent boys to mumps in the hope of avoiding complications sometimes associated with infections after puberty.

Why the four-year delay between mumps and the onset of diabetes in the Erie County youngsters? The mumps virus is not always a hit-and-run bug. In addition to the familiar painful swelling it causes in the jaw, it can also cause an acute inflammation of the pancreas which usually subsides in a reasonable time. In a few instances, however, mumps virus has been recovered from diabetic children long after recovery from the mumps.

There is a distinct possibility that the virus takes up long-time residence in the pancreas, where it can destroy the insulin-producing cells directly or alter the pancreatic tissue in such a way that the body no longer recognizes it as its own. When that happens, the body's defenses spring into action to reject it as they do a foreign transplanted heart or kidney. Further evidence of this series of events is the observation that the majority of juvenile diabetics develop antibodies against their own pancreas tissue during the first year of diabetes. Rejecting one's own tissue is called auto-immunity, an allergic reaction to your own tissues.

Meanwhile, the infection theory of diabetes continues to gain strong support from experiments with laboratory animals. At the University of Vermont, Dr. John E. Craighead, who was not looking for diabetes, found it nevertheless when he infected mice with a flu-like animal virus (EMC). When Pennsylvania State University's Dr. Bryce L. Munger placed healthy young guinea pigs in a cage with diabetic animals or injected them with urine of diabetic animals, more than 50 percent of the hitherto healthy animals developed diabetes.

In 1976, Boston University pediatrician Dr. Sidney Kibrick and Medical College of Virginia microbiologist Dr. Roger M. Loria triggered diabetes in mice with Coxsackie B — a first for infection with a known human virus. Only the genetically susceptible animals succumbed; the offspring of two diabetic parents all developed diabetes, whereas only half the mice with one diabetic parent were affected, and that at a later date. The mice with no diabetes in their heredity escaped the disease despite infection with the same virus.

In the spring of 1977 came the first proof that a virus can destroy the beta cell (the insulin-producing cell in the pancreas of humans), when a young doctoral candidate,

Thomas Stonebach, working with Dr. Richard L. Crowell, Hahneman Medical College, Philadelphia, exposed cultures of human beta cells grown in the laboratory (a research tool yielding important information in a variety of diseases) to a particular strain of Coxsackie virus. The cells were destroyed. Moreover, this particular virus has properties that may make it suitable for a vaccine in the future.

All of the viruses implicated to date are common sources of infection — mumps, German measles, Coxsackie. And the laboratory animals who have succumbed were all genetically susceptible — the product of breeding of generations of diabetic ancestors. The role of heredity in juvenile diabetes, however, has long been a puzzle.

For the most part, many juvenile diabetics like Robbie do not have a family history, and those who do often exhibit no clear-cut pattern. Tracing her roots, Karen, a diabetic since age 10, tells me her father just developed mature-onset diabetes at 43. His father at 76 still has no sign of diabetes, but her great-grandfather, who died recently, was an insulin-dependent diabetic.

Is there a genetic vulnerability of the pancreas to viral infection? If so, what is its nature? The search led Dr. Jorn Nerup of Copenhagen in a seemingly unrelated direction — research spawned by a serious problem in organ transplants. From the earliest days of transplants, one of the major barriers to success has been finding a donor kidney, heart, liver, etc., that the recipient will not reject. Only an organ from an identical twin, sharing the same genes, can be depended upon to be compatible. What the doctors did not know was the mechanism the body uses to recognize tissues of different genetic origin, and what determines compatibility.

By the time Dr. Nerup undertook his inquiry, several

dozen different factors governing compatibility had already been identified on the surface of white blood cells — HLA (human leukocyte antigens). Like all proteins that make you uniquely you, your HLA are dictated by your genes. By this time it was also known that some of the compatibility factors or antigens appear with increased frequency in both viral and auto-immune disorders, a category to which juvenile diabetes was a relative newcomer.

Dr. Nerup studied HLA patterns in a series of diabetics and by late 1974 he reported in *The Lancet* that he found the genetic markers for juvenile and other insulin-dependent diabetics — a frequency of HLA 8 and HLA 15 — two to three times as high as in the general population.[16] Mature-onset diabetics like Rosa and, as would be shown later, MODYs like Jane, who still have the capacity to make insulin, do *not* have these markers — another argument that these are indeed different disorders.

Ongoing research since then by Dr. Nerup as well as by Dr. J. C. Woodrow and his colleague, Dr. A. G. Cudworth, bears out the increased susceptibility in an individual in whom HLA 8 and HLA 15 are present. In families where two or more children are victims, each has the same HLA genetic pattern. Susceptibility to juvenile diabetes among those in whom both markers are present may be even stronger than hitherto suspected. In a letter to the *New England Journal of Medicine*, February 1977, researchers at Semmelweis University, Budapest, rate the risk as eighteenfold.

The HLA genes are located on chromosome no. 6, one of 46 chromosomes in every cell in your body. You inherit these genes in the same manner that genes for eye color or

---

[16] At the same time, but independent of Dr. Nerup, Dr. J. C. Woodrow made the same discovery in juvenile diabetics in Liverpool.

blood group are transmitted — one from each parent for each trait. Also located on chromosome no. 6 and closely linked to the HLA genes are genes that determine your immune response: how you deal with a foreign invader — bacteria, virus, foreign tissue or your own tissue — when self-recognition breaks down as it does in auto-immunity.

There is still no proof of exactly how this all adds up to cause disease, but Dr. Nerup proposes that because of an inherited altered immune response, the infecting virus succeeds in destroying the beta cell of the pancreas where insulin is made, and in some instances triggers an auto-immune reaction, causing more extensive damage to the pancreas. The outcome is a lifelong dependence on insulin for diabetics like Robbie and Karen.

More recently, an HLA marker has been identified that *protects* against juvenile diabetes — HLA 7. It may now be possible to predict not only who is genetically vulnerable but also who is better disposed to resist damage to the insulin-producing cells of the pancreas when hit by a viral infection.[17]

Will juvenile diabetes some day join the list of preventable diseases like polio? In the light of these findings, Dr. Cahill is optimistic that it can. When? Only when the vulnerable candidate can be identified easily and accurately, when the offending virus has been nailed down, and a safe and effective vaccine made. A decade ago it would all have been considered fanciful speculation!

When Robbie, Karen, Jane and Rosa's younger children are ready to raise a family, they will want to know "what is the risk?" As in all genetic counseling, their decision will

---

[17] The association between HLA factors and susceptibility to eighty other diseases is now under investigation. Included are arthritis, multiple sclerosis and glaucoma. (See page 198, Chapter 6.)

be based not only on the probability of transmitting the diabetic gene, but also on the burden imposed by living with the disorder.

Unlike some of the single-gene hereditary diseases, such as cystic fibrosis, sickle cell disease, Tay Sachs, etc., where the pattern of inheritance is understood and can be predicted, diabetes shows no such consistency. The age-old knowledge that it is a family affair still holds, but despite recent discoveries of receptor defects in mature-onset diabetes and HLA factors in juvenile diabetes, much of the genetics remains a mystery.

The most clear-cut pattern of inheritance is Jane's — she, her mother and her grandmother are all mild diabetics, and all three are doing well without insulin or other medication. If she transmits the disorder to *her* children, the odds are it will be mild and easily controlled.

Rosa's children are also aware of their vulnerability, but, like their mother, they may not know about their status until after their child-bearing days are over. In any event, they also know how to protect themselves with a lifelong regimen of sensible diet and good weight control.

By contrast, the juvenile diabetic bears a low risk of passing the disease on. In one study, 11 percent of the juvenile diabetics had a diabetic parent (some, like Karen's father, identified after the child), 8 percent had brothers or sisters with diabetes, and 6 percent were third-generation diabetics. One doctor at a recent symposium described the ultimate in the genetic puzzle — identical twins, one of whom was a juvenile diabetic and the other a mature-onset diabetic. The overall risk of juvenile diabetes in a child of Robbie's or Karen's is small, even if each marries another juvenile diabetic. On the other hand, there is a large burden imposed by the disease itself on the young mother or father. Some, like other young persons free of disease, are opting for no children.

"Pregnancy is no threat to the life of the diabetic mother," says Dr. White. There are risks, however, for the baby. Spontaneous abortion is more frequent, and when the pregnancy is successfully completed, the possibility of congenital malformation in the baby is significantly higher than in the non-diabetic mother.

Such a problem has long been recognized, but since congenital abnormalities occur in children of non-diabetic mothers as well, it has been difficult to assess how much the diabetes contributes. Now, after a study involving 50,000 mothers, half white and half non-white, the answer is here. A diabetic mother is twice as likely to have an abnormal child as a non-diabetic mother. Defects include heart problems, spinal abnormalities, severe eye disorders, and respiratory disease. For some still unexplained reason, serious abnormalities are more frequent in children of white diabetic mothers, while the children of non-white diabetic mothers suffer minor abnormalities.

At fault is an unfavorable environment in the uterus of the severe diabetic mother and is unrelated to the diabetic gene. Indeed, if Robbie marries a non-diabetic, their children will be at no increased risk of birth defects.

Pregnancy may intensify overt diabetes or precipitate a hitherto unknown problem. Dorothy learned about it during her second pregnancy. She had always been in excellent health, and everything went smoothly with Timmy, who is now 3. To be sure, Timmy was a big baby — he weighed nine and a half pounds. But Dorothy is tall and robust, as is Tim Sr.

This time it is different. She begins to show signs of diabetes during the pregnancy, but her doctor reassures her. "Most likely, after the baby is born you will have no sign of diabetes at all." What Dorothy has is a transient problem and she can be reassured that there is no increased risk of birth defect for the baby. The doctor's prediction turns out to be correct. After delivery,

Dorothy's sugar metabolism is again normal. She had what is called "gestational diabetes."

In the light of recent knowledge, however, gestational diabetes is more common than suspected and by no means a completely innocuous condition. At risk are young women who have a family history of diabetes (one of Dorothy's grandparents on each side is a diabetic); are obese; are over 25; have already had a stillbirth or an abnormally large baby. Unrecognized and untreated, "it may threaten the baby's good health, accounting for as many as 4,500 infant deaths every year," warned Case Western Reserve's Dr. Irwin R. Merkatz at the Cornell University–March of Dimes Symposium in May 1976. Moreover, 50 percent of the women with gestational diabetes will become diabetic within fifteen years.

For Dorothy, the warning is clear. She must avoid obesity and keep physically active. The changing life pattern of such women is reflected in protection not only for themselves but for their children as well. A follow-up study of children of mothers with gestational diabetes reveals that despite the hereditary predisposition to obesity, most of these children are still not obese by the age of 15.

Added to the physical, psychological and emotional cost of diabetes to the patient and his family is a dollar price — estimated at 5 billion dollars a year, even without considering the cost of complications (surgery and rehabilitation from gangrene, eye surgery, etc.). The price for a "healthy" diabetic, leading an active, normal life, is soaring. "In a very short time," says the mother of two juvenile diabetics, "insulin has gone up from $1.59 to $4.59." The insulin-dependent diabetic who is over 65 and on a fixed income may find the cost even harder to carry, since Medicare does not cover out-of-hospital medication.

On the plus side for the diabetic is the new public

awareness that this is a disorder for which no cure has yet been found, and a growing familiarity with its widespread impact on the lives of millions. Such awareness is a must if a meaningful commitment to support research is to be made.

The American Diabetes Association has long been in the forefront of education, notably for the profession, but with the soaring numbers of juvenile diabetics, a need arose for a source of help for their unique problems. Lee Ducat describes her "devastation when my 9-year-old son was struck. We desperately needed some place to go for help."

She soon found other families with the same plea, and from the formation of the first chapter in Philadelphia in 1970, the Juvenile Diabetes Foundation has grown to eighty-four across the country. "We are international now," says National President Carol Lurie, "with two chapters in Canada, one in Tel Aviv, and a beginning in Copenhagen, where parents of juvenile diabetics are getting together to share their problems and hopes."

There is hope in the action taken on the last night of the ninety-fourth Congress in October 1976, when members voted to continue the work of the National Commission on Diabetes, and to remain committed to continue a program of education and research.

How does a diabetic view the present and the future? Not too long ago, I chatted with a 29-year-old physician whom I had first met at a 1976 New York Academy of Sciences Conference on Chronic Cannabis Use. We had both listened with interest to a paper on the health status of chronic heavy users — those with a minimum of ten years of regular use three or four times a week. Extensive laboratory tests revealed no difference between users and nonusers in fasting blood sugar level or in sugar level two hours after ingesting glucose. How does this relate to a diabetic who uses marijuana? We knew of no such studies in diabetics, but the young doctor shared some anecdotal observations with me. The consensus is that blood sugar may go down, and that the diabetic would be wise to have a snack before he smokes.

This time he was speaking of more far-reaching events. "When I became a diabetic fourteen years ago, no one dreamed of a brain hormone that could help insulin; the thought that I am sick because of a virus infection would have been scoffed at; talk of lasers and vitrectomy in blind diabetics would have been more appropriate to science fiction than a medical journal. Today, with the momentum of new knowledge, how can I be anything but optimistic?"

5

Obesity in Perspective

When President Jimmy Carter visited the nuclear submarine *Los Angeles*, five crew members were enjoying an unexpected day off. All overweight, they had been relieved of active duty for the day by the captain, presumably because their obesity would not present a shipshape image.[1] At a private midwestern university, a student leader with a 3.4 gradepoint average was suspended because she could not measure up in slimming down. She is now one of a group asking an HEW review to determine if the university is guilty of discrimination.[2] Scorn for obesity is not new. Socrates' perception was that "bad men live that they may eat and drink, whereas good men eat and drink that they may live."

Socrates was wrong. Overnutrition is neither immoral nor ignoble. On the contrary, it is a frustrating and costly experience for the 30 to 50 million obese Americans who last year spent 10 billion dollars on appetite depressants, anti-obesity prescriptions, doctors' fees, diet books, spas, mechanical devices, synthetic diets, etc. By next year,

[1] *New York Times,* May 3, 1977.
[2] *New York Times,* December 4, 1977.

probably 80 percent of those who lost weight last year will have regained much of it. Their bruised self-esteem may be lower than ever, but many will nevertheless start a new cycle of what Dr. Jean Mayer[3] calls the "rhythm method of girth control."

Meanwhile, the upward trend begun with the changing lifestyle after World War I continues. A late 1977 report by the National Center for Health Statistics found that Americans are more overweight than they were a decade ago, with the average man 20 to 30 pounds too heavy, and the average woman 15 to 30 pounds too heavy. U.C.L.A.'s Dr. George A. Bray estimates that 10 to 30 percent of all Americans weigh at least 30 percent more than their "ideal" weight. It should be pointed out that obesity and overweight are not synonymous. Weight is the sum of muscles, body fluids, bones and fat. Obesity is an excessive amount of fat tissue. See the chart on page 156 for desirable weights.

Just as dismaying is the candid acknowledgment by a prestigious international panel of researchers and clinicians at an October 1977 NIH sponsored conference of their "ignorance about what causes obesity and how best to treat it."[4] It would appear that not much has changed since a study a decade earlier ruefully reported that "despite an imposing body of information derived from research, our ignorance concerning the etiology, pathogenesis, and treatment of obesity is remarkable."[5]

Nevertheless, there are today new perspectives on several fronts that may well provide a more optimistic framework, if not a panacea, for the future. Many widely

---

[3] At the time a noted nutritionist at Harvard University School of Public Health, now President of Tufts University.

[4] *Science*, December 2, 1977.

[5] *Pediatrics*, Vol. 40, 1967.

DESIRABLE WEIGHTS FOR MEN AND WOMEN
According to Height and Frame, Ages 25 and Over

HEIGHT (in Shoes) Men (in 1-inch heels)	Weight in Pounds (In Indoor Clothing)			HEIGHT (in Shoes) Women (in 2-inch heels)	Weight in Pounds (In Indoor Clothing)		
	SMALL FRAME	MEDIUM FRAME	LARGE FRAME		SMALL FRAME	MEDIUM FRAME	LARGE FRAME
5' 2"	112-120	118-129	126-141	4' 10"	92- 98	96-107	104-119
3"	115-123	121-133	129-144	11"	94-101	98-110	106-122
4"	118-126	124-136	132-148	5' 0"	96-104	101-113	109-125
5"	121-129	127-139	135-152	1"	99-107	104-116	112-128
6"	124-133	130-143	138-156	2"	102-110	107-119	115-131
7"	128-137	134-147	142-161	3"	105-113	110-122	118-134
8"	132-141	138-152	147-166	4"	108-116	113-126	121-138
9"	136-145	142-156	151-170	5"	111-119	116-130	125-142
10"	140-150	146-160	155-174	6"	114-123	120-135	129-146
11"	144-154	150-165	159-179	7"	118-127	124-139	133-150
6' 0"	148-158	154-170	164-184	8"	122-131	128-143	137-154
1"	152-162	158-175	168-189	9"	126-135	132-147	141-158
2"	156-167	162-180	173-194	10"	130-140	136-151	145-163
3"	160-171	167-185	178-199	11"	134-144	140-155	149-168
4"	164-175	172-190	182-204	6' 0"	138-148	144-159	153-173

Note: Prepared by the Metropolitan Life Insurance Company. Derived primarily from data of the *Build and Blood Pressure Study, 1959*, Society of Actuaries. Reprinted from *Statistical Bulletin* October 1977, Metropolitan Life, New York.

held beliefs about the causes and effects of obesity are now being challenged, including the concepts that obesity is always a hazard to your health; that there is a clearly defined fat personality and behavior pattern that sets the obese person apart from the lean; that fat babies become fat children who become fat adults who stay fat; that you may have been born to be fat. Not necessarily!

A growing body of evidence[6] reveals that a lifetime of

[6] From the Framingham Study, the International Cooperative Study, and data from Britain, Scandinavia, etc.

moderate obesity poses no risk to your health or longevity
if you are free of high blood pressure, diabetes and elevated blood fats (cholesterol and triglycerides) and have
not inherited a vulnerability to any of these disorders. On
the other hand, the dangers of coronary heart disease,
stroke and complications of diabetes are sufficiently serious to make weight reduction a matter of highest priority.[7]

☐ High blood pressure: it has been estimated that control
of obesity in the United States white population could
reduce hypertension (the most potent precursor to
CHD) by 50 percent. Many hypertensives who lose
weight can often get along with smaller doses of medication; some need no pills at all.

☐ Diabetes: 80 to 85 percent of mature-onset diabetics can
be controlled with diet and weight loss — a growing
practice with doctors who would like to cut down pre-
scribing anti-diabetic drugs.

☐ Coronary heart disease: while obesity by itself is no
risk, new evidence from Framingham reveals that when
it is combined with two other established risk factors —
diabetes and low HDL[8] — it carries *an especially high
CHD risk* in women over 50, and may be "equally a
matter of concern for younger women as well."[9]

☐ Gall bladder disease: a risk for the markedly obese.

☐ Arthritis and gout: aggravated by the added burden to
weight-bearing bones and joints. Pain and damage to
tissues can be relieved with weight loss. An added bonus
for gout victims is the lowering of elevated uric acid
levels.

Despite the awesome hazard of obesity when it occurs
in combination with other disorders, there are substantial
numbers of healthy (if not always happy) fat men and

[7] See Chapters 2, 3 and 4.
[8] See page 48, Chapter 2.
[9] Tavia Gordon, William P. Castelli, M.D., *et al.*, *Annals of Internal
Medicine*, October 1977.

women who live to respectable old ages, including my father, who was obese only for the last twenty-four years of his life, and who died at 91, with a normal blood pressure and no evidence of diabetes or CHD. Indeed, during the same recent period that overall average weights have been climbing in the United States, mortality rates have been declining, with an increase of three years in life expectancy for both sexes, reflecting the waning of the CHD "epidemic."

The new look at obesity denies that you get on a weight track early in life and are destined to remain there. "Many chubby infants are not obese adults," pioneering researcher Rockefeller University's Dr. Jules Hirsch told a November 1977 symposium[10] on genetics and nutrition, "and many chubby adults were not chubby infants."

In a widely quoted study (*New England Journal of Medicine*, July 1, 1976), Dr. Evan Charney reported that while babies chubby in the first six months were more likely to become obese in their twenties than normal or light-weight babies, only about one-third actually did so (36 percent versus 14 percent). Almost two-thirds became normal-weight young adults. This observation is in agreement with British studies showing that 60 to 80 percent of fat babies do not become fat children.

When Harriet and Stella met at their twentieth high school reunion, it was the first time they had seen each other since graduation. Despite the many experiences they had shared while growing up, they would probably not have recognized each other without the "Hello, my name is ____" tag. They had drifted apart when Harriet went

[10] Institute of Human Nutrition, Columbia University; National Foundation March of Dimes; November 1977.

off to college at her mother's alma mater in New England, and Stella, who had won top honors in typing and office management, started her first job as a clerk in a textile firm not too far from home. Until their senior year, Harriet was always taller and fatter, while Stella was on the thin side. Today, chubby Harriet is slim and sleek at 38, while skinny Stella, although still pretty and vivacious, is at least twenty pounds overweight.

Stella and Harriet reflect a trend observed by a number of researchers in the past — that obesity in the United States (especially among women) is more common in lower socio-economic groups than in the more affluent. Now a richly documented overall picture of the origins and trends in obesity is emerging from the 1968–70 Ten-State Nutrition Survey[11] concerned with both undernutrition and overnutrition. It covers the complete life cycle of 40,000 Americans — white, black, Puerto Rican and Mexican — from infancy to age 80, and from lower, medium and upper socio-economic levels. All were studied in family or household settings.

Whether you are lean or obese as an adult is less related to what you were as a child than to the family in which you were brought up and its lifestyle, the part of town in which you live or aspire to live, and your social and economic status. *When the social stakes are high enough, obesity can be averted or turned around.*

At virtually all ages, boys and girls from higher socio-economic white families are taller and fatter than the poorer children, with the trend continuing for boys throughout their lives. The more affluent boys become fatter men, while working-class men tend to remain lean longer. Stella's husband Eddie, a respected auto

[11] Massachusetts, New York, Michigan, Kentucky, West Virginia, South Carolina, Louisiana, Texas, California, Washington.

mechanic, is virtually as trim now as he was in high school. Harriet's husband Albert, a successful tax lawyer whom she met shortly after college, is noticeably overweight, his custom-tailored suit notwithstanding. Interestingly enough, the most affluent men are lean.

With girls, however, the story is different. All show a spurt of weight gain in adolescence, and the hitherto thin poor girl may never lose it, continuing to gain well into middle age. The more affluent girl, on the other hand, *begins to lose in late adolescence* and, like Harriet, remains slim.

Harriet's motivation became strong during her early years of college. Dates with students from the neighboring men's schools were rare for a fat girl. By her sophomore year, she was eating less and refusing rides when she could walk or bike; her gym class got the same respect as her chemistry class. With other women, the motivation is getting a job in a particular field or getting ahead in the field.

For Stella, gradual weight gain imposed no great social stigma. No one minded in the office where she worked for two years before she married Eddie. Almost every adult female in her family is overweight. Now the mother of three, she is not too dissatisfied with herself. But Stella's teenage daughter, Cathy, could be the Harriet of twenty years ago. The first in the family to apply to college, Cathy has a clear idea of how she wants to look, and much as she loves her mother, her aunts and her grandmother, theirs is not the image in her future.

Was Cathy born to be fat? It has been recognized for a long time that obesity runs in families, but how much (or how little) is genetics and how much is environment was not known. In the Ten-State Nutrition Survey, family similarities were found to be even stronger than expected. Boys and girls with two lean parents tend to be the leanest;

boys and girls with two obese parents tend to be the most obese. By *age 17, the children of obese parents are actually three times as fat as the children of lean parents.*[12] Siblings show an even stronger family resemblance. Obese children tend to have fat sisters and brothers. If one child is obese, there is a 40 percent chance that the second child in the family will be obese; and in a three-child family, there is an 80 percent chance that at least one of the other two will also be obese. Contrary to the popular belief that it is mother who sets the stage for obesity, the survey finds that one obese parent, mother or father, increases the risk for obesity in the children.

"Before suggesting that the level of fatness is primarily gene-determined, it is appropriate to consider the fatness similarities between husbands and wives," caution Drs. Stanley M. Garn and Diane C. Clark.[13] The correlation is very significant (0.3), reflecting the importance of a common style of eating. It takes time, however, for the similarity to develop — the correlation is somewhat lower among younger couples.

In a related study, Dr. Garn and his colleagues at the University of Michigan School of Public Health[14] looked at 429 parent–adopted child relationships. They found the same correlation in fatness between adopted children and parents as between biological children and parents; in fact,

[12] Obesity was determined by fatfold measurements, which are more precise than height-weight. If you think you are obese and are not sure after looking at yourself in the mirror, stripped, you can find out by taking a pinch of skin from the back of the upper arm, the side of the lower chest, or just below the shoulder. If you can grasp more than one inch, you are obese.

[13] Ad Hoc Committee to Review the Ten-State Nutrition Survey, *Pediatrics*, April 1976.

[14] Stanley M. Garn *et al. American Journal of Clinical Nutrition*, October 1976.

the former was slightly higher (0.39 versus 0.32). Furthermore, in a study of 156 families with both biological and adopted children, the correlation between the siblings was the same.

The family resemblance, moreover, extends to the family pets, as reported in a veterinarian's study of "Obesity in Pet Dogs."[15]

While no serious student of obesity doubts that heredity has a role (inherited body build, inherited patterns of laying down fat, possible enzyme and hormonal functions), recent studies cut down the genetic contribution in humans to a very small number of individuals. Because genetic obesity in animals is so well documented, it is tempting to translate the experimental laboratory observations to the human experience, but even in susceptible animals, environment and lifestyle count. The sand rat (*Psammomys obesus*), whose natural habitat is the arid desert of Egypt and Israel, and whose natural source of food is the sparse vegetation found there, only shows his genes when he is moved to the laboratory and fed a much richer diet.

"The genetic components [in humans] are there," Dr. Hirsch says, "but they are hard to tease out — there are so many other complex factors."

Becoming obese is a gradual and insidious process of taking in more calories than are used, whether it is the slim size 10 who can now barely fit into a size 16, or the rare grossly obese individual who is listed in the Guinness Book of Records. "The fattest man in Britain [740 pounds] acquired an excess of 1¾ million calories by the time he was 22," commented Dr. J. S. Garrow.[16] "He could have done it with only 218 excess calories a day."

[15] E. Mason, "Obesity in Pet Dogs," *Veterinary Record*, Vol. 86, 1970.
[16] *Post Graduate Medical Journal*, Suppl. 2, 1977.

Fat begins to accumulate in the last three months of fetal life and continues to grow for at least eighteen months, with the cells increasing in size for the first year, and in number between the ages of twelve and eighteen months.[17] Indeed, without it, independent life after birth would not be possible. Not only is fat the most efficient reservoir for energy, it is also necessary to provide the insulation that all warm-blooded creatures (including humans) require, and to provide protection against injuries to bones and internal organs. The obese all have larger fat cells — some have more, some have the normal number — both among obese children and obese adults. During periods of weight loss, the size of the cells can shrink, but the number never decreases. Furthermore, it is now known that new fat cells can be produced throughout adulthood. Whether the excess fat is contained in larger cells or in additional cells, the problems it poses are identical — how and when to prevent it, how best to get rid of it.[18]

Since most persons maintain a relatively stable body weight without measuring every calorie they consume, how do their regulatory systems differ from those of the obese? For a time, one popular hypothesis held that lean people eat in response to internal cues (they feel hunger) and stop eating when another cue tells them they are satiated. By contrast, the obese person is supposed to respond to external cues (sight of food, easy access, etc.) and to keep on eating because he has no signals that he has had enough.

Laboratory experiments demonstrate that there are such signals in animals to regulate appetite and food intake.

[17] A. Häger et al., Metabolism, 1977.
[18] Counting and measuring fat cells is still a research technique used in laboratories in the United States, England, Scandinavia, etc., and not yet a clinical or diagnostic procedure.

When Rockefeller University's Drs. Neal Miller and Eric Stone stimulated the region in the brain associated with appetite (the hypothalamus), the animals, as expected, kept eating and kept gaining weight. When no longer stimulated, however, they ate *less and less*; those who were very heavy stopped eating spontaneously until weight was stabilized. "That animal," says Dr. Miller, "was sending a signal to the brain, 'I am too heavy.' "

Evidence that such a signal exists in animals comes from the Jackson Laboratories, Bar Harbor, Maine, where biochemists created artificial Siamese twins — one of the pair a genetically obese diabetic mouse, and the other a normal mouse. With the joint circulation, the normal mouse now had a signal, "You are too fat." He stopped eating and starved to death.

Scientists believe that they have identified one such signal — glycerol — found in fat cells and also in the bloodstream. The theory, backed by accumulating evidence, holds that elevated blood glycerol gets the message through to the central nervous system (the brain and the neurohormones) that there is too much fat and it is time to change — reduce food intake, increase energy expenditure. As early as 1941, it was shown that long-term application of glycerol to the skin of rabbits prevented weight gain during the period of application.[19] Experiments in recent months and years with rats, rabbits and chickens continue to confirm that glycerol — either fed in massive amounts or given by injection — cuts down food intake and results in weight loss.

Why, then, do obese persons who have high glycerol levels maintain their elevated body weight? One possibility suggested by researchers is that the central nervous mechanism responsible for detecting glycerol may be ab-

[19] W. Deichman, *Ind. Med. Ind. Hyg.*, Sec. 2, 1941.

normally insensitive to the signal and does not read it correctly.

Can this line of research be the first step in finding the "holy grail" that will bring down obesity in humans? The problem is much more complex than a breakdown in a single biochemical marker. Animals eat enough to take care of their needs for energy, growth and reproduction. For humans, nurture and civilization have made eating a currency of social life.

No single or clear pattern of eating behavior or psychology can be attributed to all obese persons.[20] Some become depressed *after* they are obese, largely because of what they perceive as rejection. Some eat more under stress; some eat less. In fact, even under normal circumstances, a substantial number of obese consume fewer calories than their lean counterparts.

Why do they remain fat? The one characteristic that many share is *less physical activity*. When a group of six-month-olds were observed over a period of time, it turned out that the extremely thin infants were more active and ate more than the fat infants, who ate less and moved around less.[21] In obese teenagers, this behavior pattern is even more pronounced. Dr. Mayer took 30,000 pictures of obese teenagers at play at a summer camp. In a volleyball game they stood still 80 percent of the time; in the water they swam no more than 20 percent of the time; on the tennis court they spent 65 percent of the time not moving. Overall, teenage obese girls were about one-third as active as their better-proportioned friends.

How early in life should dietary intervention begin?

---

[20] Dahlem Conference on Appetite and Food Intake, Berlin, West Germany, 1976.

[21] A. E. Rose and J. Mayer, *Pediatrics*, Vol. 41, 1968.

Recognizing that optimum nutrition is a *must* in the early months of life, knowledgeable doctors also point out that optimum is not *super*-nutrition. There is a consensus that premature introduction of cereal and other solid foods can contribute to early obesity. Where early obesity is a problem, British and Swedish specialists recommend between 5 and 7 as the best age to take action, particularly if there is obesity in the family; others in the United States and abroad think puberty offers the best chance for success. True prevention, however, is a family affair and focuses not only on the potentially obese child but on an ambiance that rejects too many snack foods, too many rich foods, and too sedentary a life.

In a society where thin is in, the motivation for adults and older adolescents to lose weight for cosmetic reasons can achieve striking results. Dr. Bray has seen as much as a hundred pounds shed at a time of impending engagement or marriage. "One woman successfully dieted and became winner of a competition to be a candidate for Miss America," Dr. Bray recalls, "only to find that following the national competition, her weight gradually reaccumulated to reach its original level." The fact is that as many as 85 percent of all dieters (including those under a doctor's care) have regained the weight lost by the end of two years. Many have lost on crash diets, almost all of which are doomed to failure on a long-term basis.

Despite the forthright admissions by the most renowned researchers and clinicians about how far they are from a "cure" for obesity, panaceas and miracle diets continue to proliferate.

Most eagerly accepted are diets that promise you *all you can eat* while you lose. More than a hundred years ago, Britain's Dr. William Harvey devised a diet containing large amounts of meat (protein) and prohibiting sweet and starchy foods (carbohydrates). His patient, William Banting, later published a testimonial to its success in "A Letter of Corpulence Addressed to the Public." Versions of this diet continue to be rediscovered, and to be embraced enthusiastically before the inevitable disenchantment sets in. Early in the century, "Eat and Grow Thin: the Mahdah Menus," had gone through 112 printings by 1914. The momentum picked up in the fifties with A. W. Pennington's "Treatment of Obesity with Calorically Unrestricted Diets" in 1953; "The Air Force Diet" (no connection with any nation's air force) in 1960; Dr. Herman Taller's *Calories Don't Count* in 1961; "The Drinking Man's Diet" in 1964; Dr. Irwin Stillman's *The Doctor's Quick Weight Loss Diet* in 1967; and Dr. Robert Atkins's *Diet Revolution: The High Calorie Way to Stay Thin Forever* in 1972.

The most recent popular arrival in the area of "eat and grow thin" diets is the Scarsdale regimen, which has spread by word of mouth and photocopies of the week's menus. Men and women who have lost significant amounts of weight (not for the first time) are eagerly sharing their experiences with admiring friends in locker rooms, commuter trains, beauty shops and club meetings. To date, the success of most of the advocates lasted no more than a few months. The real test will be in two years. Will they be among the 20 percent who keep the lost weight off, or among the 80 percent who are back to where they were before "the best diet yet" came into their lives?

Millions of books have been sold and thousands of patients have lined up in the respective doctors' offices, and with good reason — the diets all work, but not for long. Many find it difficult to adhere permanently to the monotony and poor palatability of the diets. Others find the side effects — especially ketosis (acidosis associated with abnormal breakdown of fat tissue) — intolerable.

Actually, many persons eat fewer calories than before
— 13 to 55 percent fewer. "How much butter can you eat
without bread?" one devoté asks. As for the need to eliminate virtually all carbohydrates, the leanest populations in
the world got that way on a *very high carbohydrate* but *low
fat* diet (rice, vegetables, etc.). Moreover, there is no evidence that a high-calorie diet limited to fats and proteins is
less fattening. It may be more filling. It may also be more
hazardous.

In 1976 a new diet with a special appeal to defeated
veterans of other campaigns swept the country — Dr.
Robert Linn's "Last Chance Diet." In contrast to the "eat
all you can" regimen, the last chance is a modified fast (as
low as 300 calories a day), with food intake limited to a
predigested liquid-protein product derived from animal
hides and other tissues.[22] "These products have simply
taken the United States and Canada by storm," FDA
official Dr. Allan L. Forbes told the October 1977 NIH
International Obesity Conference and alerted the participants that "the FDA has received reports of illness and
deaths of persons alleged to have been subsisting primarily
on these products for periods from weeks to several months."

By early 1978, the FDA and Center for Disease Control
(United States Public Health Service) were investigating 58
deaths associated with the liquid protein diet, of which 16
fit a distinct clinical and pathological pattern.

All were women between 25 and 51 who had dieted an
average of five months and had lost an average of 83
pounds. The majority had been under strict medical

---

[22] The concept of sharply restricting calories and limiting the small
food intake to proteins has been applied successfully since 1973 by
Boston's Dr. George L. Blackburn, whose aim is to achieve weight
loss by burning only fat and sparing body protein. Dr. Blackburn
uses meat, egg albumin and other high-protein *foods*.

supervision, 14 took vitamins and minerals, and 13 took supplemental potassium. All had been feeling generally well — and all had died suddenly and unexpectedly, some while still on the diet, others within a few weeks after they went off it. Immediate cause of death: derangement of heart rhythm that could not be corrected by drugs, electrolyte therapy or direct countershock to the heart.

Not surprisingly, sales of the liquid protein (about fifty brands) fell sharply after the public disclosure of its possible hazards. Even before the deaths, doctors on an ad hoc advisory committee to the FDA, most of whom had had experience with the diet, described some of the side effects — nausea, vomiting, diarrhea, constipation, muscle cramps, blackouts, sharp drop in blood pressure upon standing up, recurrence of gout, and others.

While it is still not possible on the basis of current evidence to say definitely that any of the 16 deaths were caused by adherence to the diet, it is now known that the risk of death for women aged 25–44 who use it for more than two months is almost 30 times as great as for women in that age group in the general population. An editorial in the May 4, 1978 *New England Journal of Medicine* suggests that ". . . the liquid protein diet regimen should be used with extreme caution and only under carefully controlled conditions. . . . *conventional medical supervision is not an adequate safeguard*." (Authors' italics.) "Additionally," warns CDC in a July 7, 1978 report, "prolonged use should be limited to research settings controlled by protocols approved by committees on human experimentation, and only with the informed consent of the participants."

We know one young woman who needed to lose many pounds because (1) her mother is a diabetic; (2) she plans to go back to work now that her youngest child is in school; (3) she is not happy with what she sees in the mirror. She lost 55 pounds in six months under a doctor's supervision.

Today, six months later, she has regained 45. At her weight, the net loss of 10 pounds is not significant. She could have lost that amount without fasting at all — but by walking a mile a day every day during the past year.

Who are the 20 percent who have lost weight and kept it off? Like Harriet, most are first-time losers who continue to maintain constant vigilance. "It has not always been easy," Harriet admits, recalling the periods after the birth of each child, and the first few months she was on the Pill. She still has menopause ahead, a time when many women gain, partly because it often coincides with a decrease in physical activity. Losing weight on essentially the same type of diet that will be followed for a lifetime is an important part of a successful program. The diet should be varied, balanced, palatable, and should, by all means, include some of your favorite foods — in judicious amounts, of course. Nothing threatens a diet as much as self-pity and a feeling of deprivation. Commitment to more physical activity both for fun and for daily living is a must — stairs instead of elevators if it is only a flight or two, errands on foot or by bicycle if possible, etc. Most important is continued support (not sermons or lectures) from family, friends and co-workers.

Such organizations as TOPS (Take Off Pounds Sensibly) and Weight Watchers, which provide group support, are attracting hundreds of thousands. While the dropout rates are high in both programs,[23] for those who remain the results are achieved safely and sensibly, and they know that they can always rejoin the group when they feel they need it.

Behavior modification, while promising, is still in the early stages. Quipped a speaker at a recent nutrition meet-

[23] Follow-up is limited.

ing when asked if behavior modification has a role in obesity, "Of course — it brought it on."

Until not too long ago, teaching nutrition to medical students had an extremely low priority. That is changing now and should, in the future, be reflected not only in prevention of obesity, but in management as well. A recent survey[24] finds that 19 percent of medical schools now require a course in nutrition and 70 percent offer it as an elective. It is significant that increased student interest was frequently the stimulus for including nutrition in the curriculum. There are still difficulties to be overcome — lack of funds, lack of time in the curriculum, lack of qualified faculty.

Finally, the new look at obesity is beginning to question whether everyone who is trying to reduce should. "Leave alone those who are 15 to 20 percent over," pleads a psychiatrist who has long been associated with the problem.

Can you leave it alone? Probably only when the acceptable dimensions are closer to a Rubens than a Modigliani.

[24] Cathy Cyborski, M.S., *Journal of the American Medical Association*, June 20, 1977.

6

Aging — A Wreath of Thorns?

Our friend Walter is back at college, enrolled in philosophy courses he could not fit into his undergraduate program more than half a century ago. At 79, he is studying Hume and Huxley, doing his homework meticulously, and finding that "the time we spend in class goes by so quickly, it's amazing." Describing the growing graying generation on campus as "magnificent students," a professor at Walter's college adds: "They bring a four-letter word to each classroom they are in — LIFE."

Shortly before his eighty-seventh birthday, our Uncle Louie, an ex-carpenter, wood-paneled his kitchen. "I only did it waist high," he explains, but he is clearly satisfied that he has not yet lost his skill and strength.

Ann, at 81, "still has the spirit of joy in her life," one of her many friends and admirers tells us. It was that spirit that was to sustain her when she became deaf at 19, and that was to catapult her to national prominence for her role in establishing the first government-agency employment programs for the handicapped. Today, five years after her husband's death, she continues to maintain herself in her own home and to acquire new friends and new interests, including an almost perfect attendance at an exercise class.

Walter, Uncle Louie and Ann have all achieved the goal of gerontology[1] today — to defer aging by stretching out the health and vigor of the youthful and middle years. And in the light of today's knowledge, delaying premature deterioration and senility to the limit of our life span is a realistic possibility. There is a growing consensus that aging and disease do not necessarily go hand in hand. The healthy heart is still a strong muscle at 70; the healthy liver of a 70-year-old metabolizes alcohol as well as that of a 30-year-old; not all aging eyes cloud up with cataracts; sexual desire and activity continue; and the much-feared loss of intellect is not inevitable.

Indeed, many 70-year-olds score higher on vocabulary and general information tests than their college-age grandchildren. It may take them longer to master new tasks and knowledge, but, on the other hand, they have greater patience with tedious tasks, especially when the motivation is strong. So it is with noted journalist I. F. Stone at 70, who, on retiring after fifty years of editing a unique weekly, "fell in love with the Athens of the fourth and fifth century B.C." He has taught himself Greek because he "could not make valid political inferences [of Plato] from a translation, no matter how competent."[2]

The optimistic feeling that we can all "die young at an old age" is based on new insights and perspectives in the biology and genetics of aging. Researchers are beginning to bring in answers to age-old questions: Is our best route to longevity to be lucky enough to be born to long-lived parents? Do we live and die according to a biologic clock? If so, where is it located, what makes it tick, and can it be controlled?

The urgency to find and implement the right answers is

[1] Scientific study of aging.
[2] *New York Times*, January 22, 1978.

underscored by the fact that for the first time in history, the elderly constitute a significant proportion of the population.[3] At the turn of the century, one in twenty Americans was over 65; today, it is one in nine. By the time today's schoolchildren and adolescents are 65, they will be living in a society where they will account for one in five. Indeed, the younger you are today, the greater is the probability that you will live longer than your parents and grandparents. Whether you will sit around therapeutically tranquilized, listening to the music of your past (rock, jazz, country-western), or continue to use the intellectual, physical and sexual vigor you will doubtless retain may well depend on how the problems of aging are handled today.

The dismal fact is that Walter, Uncle Louie and Ann are not typical. Their good old age is possible only because they are in *good health*, *have what to live for* and *what to live on*. For millions, however, the "crown of roses" of youth has been replaced by a "wreath of thorns": about one-third live below or near the poverty line. Just as serious is the negative image of the over-65, based on the myths of uselessness, frailty, intellectual decline and dependence. Even more serious is what happens when they themselves accept the role, and the myth becomes reality.

"We must get away from the stereotype," warns Dr. Robert N. Butler, first director of the newly established National Institute on Aging (NIA), the government's first serious commitment to supporting research in the field.

---

[3] Arbitrarily defined by retirement age: 70 in Sweden, 60 for men and 55 for women in the U.S.S.R. In the United States, workers may retire at 65, but under a new law enacted in April 1978, mandatory retirement has been raised to 70 for most positions. Leading the successful campaign in Congress was 77-year-old Florida Representative Claude Pepper.

"Our popular attitudes towards aging could be summed up as a combination of wishful thinking and stark terror." Dr. Butler's appointment gives the elderly an articulate and dedicated advocate in government. A soft-spoken man with a warm smile, he expresses his anger at the plight of the millions of aging not in shrill complaints, but in realistic solutions.[4] For the first time in history, some of the solutions will come from the increasing number of the over-65 who are wearing the "thorns" reluctantly and resentfully. Too numerous to remain invisible and too vocal to be ignored, they are, for the first time, taking a hand in their own destiny.

"There has never been a shortage of theories about aging," says Dr. Nathan W. Shock, dean of American gerontologists,[5] "only a shortage of evidence to back them up." In the last five or ten years, however, with the development of molecular biology and sophisticated technology, it has become possible to test out old ideas (discarding some along the way) while pursuing new leads. "We no longer expect to find a general overall theory that will explain all diseases," Dr. Shock points out. "Similarly, I doubt whether a single theory of aging will explain all aspects of aging." A rundown on the many fruitful directions research is taking today attests to how correct he is.

The familiar advice to choose long-lived parents if you want a long life is only partially true.[6] To be sure, you will

[4] *Why Survive? Being Old in America* (New York: Harper & Row, 1975). The day Dr. Butler began his new job — May 3, 1976 — was also the day he was awarded the Pulitzer prize for the work that is a must for anyone who cares.

[5] Former director of the Gerontology Research Center, Baltimore, for thirty-five years.

[6] A survey by the Kiev Soviet Institute of Gerontology of 27,000

have a head start at birth with a legacy free of genes for premature coronary heart disease, high blood pressure, and diabetes, but you are by no means guaranteed a permanent ride on the longevity track. Your lifestyle and environment can wipe out that advantage. For those with a hereditary handicap, reinforced by a lifestyle that nourishes it, aging comes early. Illness from CHD, high blood pressure, and diabetes shows up before 60. Beyond that age, genetics plays a minor role in the incidence of these disorders that ultimately account for more than 50 percent of all deaths.

While you are not genetically programmed to die of a heart attack at 55 or to blow out the candles at your one-hundredth birthday party, you *are*, like members of all animal species, genetically programmed for a maximum life span. For humans, it may be as high as 120-130 (with few of us getting anywhere near that); for the Galapagos turtle, 160 years; and for the adult mayfly, one day.

Some species are programmed to die soon after reproduction. But even among these species, genes need not always be destiny. The female octopus lays two to three hundred eggs, watches them in the water, stops eating, and then dies. When Brandeis University's Dr. Harvey Wodinsky removed her optic gland, she continued to eat after spawning and lived up to nine months longer — twice her normal span. Similarly, the Pacific salmon can stave off death after spawning when the adrenal gland is removed. Dramatic extension of life span in rats was achieved more than half a century ago by Cornell University's Dr. Clive McCay, through drastic reduction in the rats' intake of calories and proteins early in life. Later, the same successes were scored with silkworms, bees and

---

persons over 80 found that only 28 to 40 percent had blood relatives who lived to extreme old age. At least one long-term study in the United States (Duke University) shows very little correlation of the 80-year-olds and the time of death of close relatives.

chickens. Dr. McCay's rats, however, did not live a *good longer* life — their overall development was retarded, their growth was stunted, and they were susceptible to infections. Translating these results to humans, a researcher recently asked, "Would you want a small, sterile grandson with a running nose?"

The case against overnutrition remains strong, however, and scientists are now extending life span in animals without retarding growth. Older rats live longer when protein in the diet is cut 12 percent, and rotifers (cold-blooded aquatic animals the size of a pin head) whose diet is switched after maturity live 20 to 30 percent longer.

Life span has also been lengthened by lowering temperature; rotifers who normally live eighteen days in 35°C water live thirty-four days when the water is cooled to 25°C. The human thermostat, however, has a much smaller range. Even a two- to three-degree lowering of temperature causes chills, discomfort, sluggishness, and mental impairment. A greater drop can be fatal. At a December 1977 special press briefing, Dr. Butler emphasized the danger of freezing to death not in a subzero blizzard (most persons take steps to come in out of the cold), but indoors, where the temperature is under 65°. The elderly are especially vulnerable to accidental hypothermia, a condition where body temperature falls below 95° and that can be fatal if not detected early enough and treated properly.[7]

There is no longer any mystery about how to provide added longevity to the 50 percent of the United States population suffering from heart disease, high blood pressure, and diabetes.[8] Now, for the first time, scientists are

---

[7] Despite the widespread pleas to lower thermostats to save energy, any room occupied by the elderly should be kept *at or above* 65° at all times. The Federal Energy Commission concurs.

[8] See Chapters 2, 3 and 4.

beginning to close in on how to delay aging and death for the other 50 percent. "The most immediate hope for significantly slowing down the rate of aging . . . lies with manipulation of the immune system by one means or another," says U.C.L.A.'s Dr. Roy L. Walford, one of the earliest and strongest proponents to recognize its importance. For with the breakdown of the immune system that comes with aging, the body loses its ability to fight bacterial and viral infections, to resist cancer, and to differentiate between its own and foreign tissues — giving rise to the formation of auto-antibodies that seek out as their targets not the foreign invader, but the body's own tissues. Long associated with arthritis, anemia, thyroid disease, etc., auto-immunity is now implicated in heart disease, hypertension and other degenerative diseases of aging. The challenge is to find a way to bolster the faltering immune system.

Although immunology dates back to Pasteur a century ago, it was only in the 1950s that a familiar white blood cell, the lymphocyte, was identified as a major source of immunity. It was not until a decade ago that it was learned that two types of lymphocytes are involved — B cells, which are responsible for antibodies circulating in the bloodstream, and T cells, which play a major role in fighting bacteria, viruses and cancer, as well as in maintaining self-recognition. Both cells have a common ancestor — a stem cell made in the bone marrow. Playing an essential role in this elegant security system is a small, hitherto obscure gland high up in the chest — the thymus.

The B cell-to-be moves out directly to the lymphatic system, while the T cell-to-be migrates to the thymus, from which it emerges with immunocompetence, not only for its own unique functions, but as a helper to the B cell as well. In the past, one of the most baffling aspects of the thymus was why it is largest early in life and then begins to shrink. Scientists now know that the shrinking thymus is also accompanied by fewer T cells, fewer antibodies against foreign threats, and more auto-antibodies against self.

In 1975, an NIH team headed by Dr. Takashi Makinodan turned the clock back for aging mice by transplanting thymuses and bone marrow from young mice. The *immune systems of nineteen-month-old mice were as youthful as those of four-month-old mice* and remained that way for six months. For humans, the boost would be equivalent to *fifteen to twenty years*. Because T cells can be frozen for long periods, then warmed up and reactivated, Dr. Makinodan[9] speculated at a December 1977 symposium on the biochemistry of aging, sponsored by the Intra-Science Research Foundation, that you might be able to bank your own T cells in your youth and withdraw them for use in your old age.[10]

The protective power of the thymus is not limited to processing T cells. In 1965, Dr. Allan Goldstein, a young biochemist at Yeshiva University, collaborating with his professor, Dr. Abraham White, isolated a substance from calf thymuses that they identified as a hormone; and they later found the substance capable of stimulating the formation of T cells in humans. After moving to the University of Texas, Galveston, Dr. Goldstein continued his research on thymosin with his new team, and their subsequent discoveries are now saving the lives of children born with such severe immuno-deficiencies that even chickenpox can be lethal, and stimulating new T cell formation in cancer patients. Perhaps one day they will be able to do the same for the aging.

[9] Now at Veterans' Administration Geriatric Research, Education and Clinical Center, Los Angeles.
[10] *Chemical & Engineering News*, American Chemical Society, December 19, 1977.

In an update of his work presented at the December 1977 symposium, Dr. Goldstein reported that twenty children are now under treatment. The first patient, on it since 1974, is relatively healthy and in school. Before the treatment, she spent the first five years of her life under constant hospital surveillance. The cancer studies are in progress in twelve cancer centers in the United States and Europe. And now Dr. Goldstein has isolated a new factor — thymosin alpha 1 — which is a thousand times as active as the substance he is using in the current tests. If scientists at the Roche Institute of Molecular Biology who are trying to synthesize it succeed, and if the promise of thymosin holds up, the next step might well be a product that will slow down aging.

A growing number of scientists today think that the secrets of aging will be found in the intricate endocrine system and in the brain. It had long been believed that the biological clock for menopause was the time when there were no more eggs in the ovary. When, however, researchers transplanted ovaries from aging rats into young ones, the estrus cycle began again. Furthermore, young ovaries transplanted to old rats did *not* cycle, suggesting that the loss of eggs is not the sole factor in declining fertility. About nine years ago, a Michigan State University team headed by Dr. Joseph Maites suspected that it might be the result of a deficiency of catecholamines,[11] the chemical messengers in the brain that carry the signal from the hypothalamus to the pituitary to trigger the ovary. Working with old rats, they succeeded in starting the estrus cycle, with the animals in "heat" again, after administering epinephrine and L-dopa. More recently the researchers measured production of these chemicals in

[11] Substances related to adrenalin and concentrated in certain nerve tracts of the brain, including the hypothalamus.

young and old rats and found a much more rapid output in three- to four-month-olds than in the aged of twenty-one months.

To date, no mammal has been found with a genetically programmed "self-destruct" timetable like that of the octopus and her optic gland. Now, however, a less dramatic but (if correct) no less effective mechanism has been proposed by Dr. W. Donner Denckla,[12] of the Roche Institute of Molecular Biology. Investigating the apparent decrease in thyroid activity (essential for good cell metabolism, immunity, etc.) with age, he found that while there was no shortage of thyroid hormones, the tissues were not responding. Apparently the receptors to respond to the hormones are deficient or defective. (Not unlike the situation of the mature-onset diabetic who is making plenty of insulin that cannot bind on to the tissue.) This explains why turn-of-the-century efforts at rejuvenation with thyroid failed.

The next break came when he discovered a new pituitary factor, emerging at puberty in the rat, dampening the receptivity to thyroid and continuing to do so until winding down is complete and death ensues. If Dr. Denckla is correct, the program for death starts fairly early in life; and if there is an aging hormone, why not an "anti-aging hormone" to counter it?

While the new look of gerontology holds that the goal of a good old age for all will be attained by learning how to control the master systems — endocrine, nervous, circulatory, immune — research on the basic unit of life — the cell — continues. For many years it was believed that while organs and systems wind down and wear out, indi-

---

[12] "A Time to Do," *Life Sciences*, Vol. 16 (Elmsford, N.Y.: Pergamon Press, 1975).

vidual cells grown outside the body are immortal. So it seemed with the chicken-heart cells that Rockefeller University's Dr. Alexis Carrel started to grow in 1912, and that were still dividing and growing in 1934. Actually, it turned out that over the years, new cells were inadvertently being introduced into the culture with the nutrients and the 1912 ancestors had long since died.

Evidence that human cells undergo a limited number of divisions (fewer with advancing age) and then die came originally from Dr. Leonard Hayflick, who reported in 1960 that cells from embryos can divide fifty to sixty times before dying, while cells from an adult double twenty-five to thirty times. Moreover, the cell is programmed with a memory. Freeze an embryo cell after twenty doublings, then thaw it, and it will continue to double thirty more times. Recently, Dr. Hayflick found that the clock is in the nucleus of the cell and that the memory persists even when an old nucleus is placed into the cytoplasm of a young cell. The number of doublings is limited by how many the old nucleus still has left. A young nucleus in an old cytoplasm, on the other hand, confers extended life to the old cell.

Nature has two models of premature aging; both, fortunately, are very rare. Progeria (Hutchinson-Gilford Syndrome) shows up early in life, and the victim usually dies in the teens, old, wrinkled and gray. With intelligence not affected, the affliction is even harder to bear. Werner's Syndrome, a genetic disorder that occurs later in life, involves early gray hair and baldness, cataracts and proneness to diabetes, atherosclerosis and cancer. They do not survive beyond the forties or fifties. In studies of progeria, it was difficult to get the cells to divide at all in culture. As for the Werners, where twenty to forty doublings were expected, no more than two to ten actually took place.[13]

[13] Recent studies on diabetics and persons genetically prone to diabetes

At the other extreme, there are cells that seem to go on forever — as immortal as Dr. Carrel thought his chicken-heart cultures were. Which are the immortals? Cancer cells!

Interestingly enough, it is not the mortality of individual cells that accounts for aging in organs and tissues. That event, Dr. Hayflick explains, takes place long before individual cells lose the programs that keep them dividing.

Do cells age because of an accumulation of errors in the DNA — the molecule in which the genetic information is stored — or does its copying mechanism become faulty?[14] Neither, apparently. First, we have a splendid repair mechanism. Also, each cell has an impressive back-up and redundancy in unused genes. At any one time, no more than 1/500 of all genes in the cell are active, and some scientists believe that this redundancy makes our longevity as a species possible. Furthermore, the DNA of aging cells continues to transmit correct information, the message gets through correctly, and the cell's machinery continues to assemble correct enzymes and other proteins necessary to maintain life. But with aging, a number of enzymes themselves show marked changes in behavior — some losing 30–70 percent of their activity between birth and death. Research, until now, has been done with rats, mice[15] and tiny worms.[16] Replacement of deficient enzymes in humans is still in the early experimental stage, but it has been successfully accomplished in at least two rare genetic disorders.

reveal a shorter cell life span with aging, in contrast to controls with *no* diabetes. Indeed, it may be a genetic marker for this continuing enigma, and additional evidence that when individual cells lose their capacity to keep dividing, it is not due to age alone — rather, to a combination of age and disease. Drs. Samuel Goldstein, J. Stuart Soeldner, *et al.*, *Science*, February 17, 1978.

[14] See Chapter 1.

[15] Dr. David Gershon, Technion, Israel.

[16] Dr. Morton Rothstein, SUNY at Buffalo.

Industrial chemists have long been familiar with free radicals — the short-lived, highly reactive atoms and fragments of molecules commonly found in rancid butter, deteriorated leather, smog, gasoline, etc. Free radicals are also found in small quantities and for brief periods in the human body, where they seem to serve no useful function. On the contrary — dedicated gerontologist, University of Nebraska's Dr. Denham Harman, proposed as far back as the 1950s that free radicals containing oxygen can inflict biologic damage on cells and promote aging. Taking a cue from industry, where anti-oxidants are used to cut down spoilage of food, paints and other products, Dr. Harman succeeded in prolonging the average life span of mice by as much as 50 percent by adding anti-oxidants to their diet. No such testing has been done with humans, and Dr. Harman believes the time has come for such a study.

While the free radical hypothesis has mixed support among other researchers, the general public has already enrolled itself in such a spontaneous, uncontrolled venture, by the enthusiastic use of an anti-oxidant that has prolonged life in fruit flies, mice and nematodes — vitamin E.[17] More than one skeptical scientist we know also uses vitamin E — "It can't hurt."

At the University of Michigan Institute of Gerontology, four researchers conducted a novel experiment in which they themselves were the subjects. To gain a better insight into how the elderly see, hear, smell, taste and feel the world around them, they spent one hour a day for six months "getting inside an older skin."[18] During that

[17] Vitamin E added to cell cultures of human embryo produced 120 divisions, in contrast to the 50 expected. But the cell died anyway. The amount of vitamin E added was no larger than the quantity normally present in serum in vivo.

[18] *Aging Tomorrow*, Vol. 1, September-October 1976.

period, they used coated lenses to distort vision, ear plugs to block hearing, a masking device to diminish smell and taste, and a fixative on their fingers to desensitize touch. The settings where they were "old" included a house, a supermarket and a senior citizens center.

Hearing impairments included a loss in the high-frequency range and difficulty in locating or identifying sounds. The researchers observed that background music interfered with ordinary conversation. Smell and taste, both for good and bad sensations, were down. As for touch, subtle differences in textures were not felt. The most marked changes they experienced, however, were in their vision. They were bothered by glare, fading colors, difficulty in judging distances due to reduced depth perception, slow adjustment from light to dark and vice versa. Names on doors and signs in public buildings presented particular problems.[19] The researchers could not, however, simulate the hardships imposed by eyes damaged over the years by high blood pressure, atherosclerosis, diabetes, weakening of eye muscles, and other effects of aging.

A recent Gallup poll revealed that the medical mishap feared most by the majority of Americans is loss of vision. Anxiety among the elderly has always been high; this is understandable considering that although the over-65 comprise about 10 percent of the population, they account for two-thirds of all persons with major visual impairment. Indeed, of the four leading causes of blindness in adults — cataracts, glaucoma, macular degeneration and diabetic retinopathy — the first three are heavily concentrated in the elderly. For a very large number, however,

---

[19] Findings of the Empathetic Model have been used in evaluating and planning changes in lighting, color coding and design in housing projects, nursing homes and hospitals.

such fears and anxieties may be groundless. The fact is that today, much of the impairment which afflicts them can be corrected or controlled.

□ Cataracts: when your lens loses its clarity and a fog descends. Although there is still no medical treatment, advances in surgery in the last thirty years have made it probably *the most successful of all surgical procedures.* If you or someone you care about is facing cataract surgery (only 15 percent of the 3 million affected will suffer enough disruption in their daily routine to warrant surgery), you will soon learn that it is accompanied by a minimum of discomfort and disablement, thanks to superior instruments and sutures, and to increasing precision made possible by microsurgery, use of a freezing probe to extract the lens, etc. You will probably be in the operating room no more than thirty-five to forty minutes, off the bed the same day (some patients walk out of the OR), and out of the hospital in two to four days. Probability of a successful outcome is 95 percent. Furthermore, age is no barrier. Many a 90-year-old is enjoying "new" vision today after cataract surgery.

□ Glaucoma: slow and sneaky. If you are at risk because you are over 38 or have a family history, your doctor can find out before the threat to your sight is serious. He or she can prevent further damage with existing medication painlessly and effectively, and can promise you even better new developments not too far off in the future.

□ Diabetic retinopathy: the fastest-growing cause of blindness in adults in the United States today, paralleling the increased life span of diabetics. Risk of blindness can now be reduced 60 percent by photocoagulation, using a green argon laser or a white xenon arc. Some diabetics are enjoying restored sight with a bold new surgical procedure — vitrectomy. Read more about it in

Chapter 4, Diabetes — New Light on an Ancient Mystery.

☐ Macular degeneration: gradual loss of central vision due to deterioration of a very small region of the retina, occupying no more than 1/100 of the entire area, but responsible for the fine vision required for reading, sewing, etc. There is still no cure, but the cause is now thought to be the breakdown of small blood vessels near the macula.[20] Lasers have worked for some, and research in their use continues. Meanwhile, the highly motivated can (1) make the most of what's left of central vision by using low-vision aids and bright lights and moving in as close as two feet to the color TV screen; (2) make the most of the intact vision of side and surrounding areas that they still retain.

The first large-scale attempt to link the development of cataracts, glaucoma, macular degeneration and diabetic retinopathy in middle and old age with biological events earlier in life is beginning to yield clues. The Framingham Eye Study,[21] a four-year cooperative investigation by the National Eye Institute and Boston University School of Medicine, performed eye examinations on more than 2,500 participants in the Framingham Heart Study[22] for whom there was a twenty-five-year accumulation of information on the state of their health and risk factors for CHD. The hope is that from the study will emerge a profile of the person prone to one or more of these eye disorders — and with identification early enough, work

[20] Demonstrated with a sophisticated diagnostic technique — fluorescein angiography — where the dye is injected into the bloodstream and its course followed to the blood supply of the retina.
[21] Headed by Harold A. Kahn, now at Johns Hopkins, and Dr. Howard M. Leibowitz, Chairman of the Department of Ophthalmology, Boston University.
[22] See Chapter 2.

out a program of prevention comparable to what is already happening with CHD.

Meanwhile, the epidemiologists found a striking similarity in risk factors associated with macular degeneration and those associated with cataracts. The findings are too preliminary to get excited about — they require corroboration and replication by other investigators — but they do point to a direction for future research.

My mother was almost 70 when the doctor told her that the "fog" that clouded her vision was not in her glasses but in her eyes. And while she was disappointed that a new pair of glasses would not solve her problem, she was comforted by his reassurance that "cataracts are the one eye disorder we are reasonably sure we can handle to your satisfaction." I was close to 40, however, when the same words and reassurances were directed to me, only a few short years later.[23] My reaction was one of dismay and disbelief — a "this can't happen to me" feeling. Cataracts, I told myself, are a disease of the elderly. The doctor's advice to "be patient and don't worry" was hard to take when I contemplated the prospect of continuing to function for years with an increasing haze over both eyes. And what would life be like after the cataracts were removed?

The events of the next few years not only cleared my clouded vision, they also dispelled many myths and mis-

---

[23] It has been known for almost a hundred years that some families are prone to the disorder, with as many as five successive generations affected. While it is still not possible to nail down a specific genetic factor, clues come from studies on identical twins 85 percent of whom have identical lenses as well as identical cataracts when they develop. (By contrast, only 27 percent of non-identical twins have identical lenses.) Complicating the task of predicting a hereditary pattern is the frequency with which opacity of the lens occurs in persons over 50, with or without a family history of cataracts.

conceptions about what it means when the clear trans-
parent lens in your eye becomes opaque, and what it means
to have the opacity removed.[24] I soon learned it would not
be necessary to wait many years for the cataracts to
"ripen," thanks to improved surgical techniques. When I
could no longer drive with confidence, read with ease, and
recognize the second baseman at the end of the block as my
8-year-old son, it was time to set the date. Recovery from
the surgery was smooth and rapid. Adjustment to seeing
with my "new" eyes was neither! It took patience and
faith — not easy for a relatively impatient skeptic.

To understand the difficulty in adjusting after surgery,
it is important to understand the role of the lens in vision.
Your eye has two focusing structures. The cornea, set into
the front of the eye like a beveled watch glass, provides
two-thirds of the focusing power. A short distance behind
it in the eye is the crystal-clear lens, less powerful, but
with a capacity (unique to humans) to change its shape and
so bring into sharp focus objects from six inches to twenty
feet away. While the cornea retains its focusing power
unchanged throughout your lifetime, the resiliency of
your lens begins to deteriorate as you approach middle
age. The eye of the needle seems smaller, and the printed
page blurred. Indeed, so commonplace is this decline after
the age of 40 that it is generally accepted as normal aging.
The remedy for presbyopia is simple and effective —
reading glasses.

[24] More than half a million men and women like me are first identified
in their forties and fifties with "senile cataracts." Others, in all ages,
may be associated with injury, excessive radiation, infection and
metabolic disorders. Diabetics are four to six times as likely to be
operated on for a first cataract at an early age than non-diabetics.
Before the immunization program for rubella, tens of thousands of
babies were born with cataracts from infection in utero. Congenital
cataracts are also associated with genetic disorders, including galactosemia.

The aging lens also undergoes more serious and more far-reaching changes. Despite its crystal-clear transparency, it is not pure water. Proteins and other substances account for almost one-third of its composition. As long as the proteins are evenly dispersed throughout the entire lens, light continues to be transmitted as effectively as if it were pure water. Changes in the protein (starting as early as the age of 20), however, will cause light to scatter rather than be transmitted. Other chemical changes also contribute to this opacity, which may result in impairment of vision ranging from a dim fogginess to total blindness.

Where you live also makes a difference. There are more cataracts in sun-baked Israel and India than in England, and a recent NEI study based on United States Weather Service maps of annual sunlight hours reveals significantly more cataracts among the over-65 who live in sunnier regions of the country.

Once a cataract has developed, it cannot yet be cured by eye drops, change in diet, or change in lifestyle. Only surgery can remove it.[25]

In the early 1970s, Dr. Charles Kelman introduced a new surgical procedure — phacoemulsification — a process that uses high-frequency ultrasonic vibration (40,000 per second) to fragment and emulsify the lens, which is then removed by suction. The incision is very small, and recuperation is rapid. Among Dr. Kelman's patients were prominent personalities who publicized their experiences widely on TV, in newspapers, etc. No longer experimental today, the Kelman procedure has been performed on more than 80,000 persons. Because it has generated such widespread public interest, particularly among the elderly, we digress here to explain what it can mean to you.

[25] See page 131, Chapter 4, for developments in preventing cataracts in diabetic animals.

Because the technique represents a marked departure from existing methods already producing superb results (complete retraining of the surgeon is required, as well as expensive equipment), a special committee of the American Academy of Ophthalmology and Otolaryngology conducted a study in 1974 to learn if its benefits are so superior as to warrant the changeover. In a personal communication, Dr. Richard C. Troutman of SUNY Downstate Medical Center, the head of the committee, wrote: "The findings of the AAOO study indicate no clear advantage of the phacoemulsification procedure over intracapsular cataract extraction as performed by the majority of the surgeons analyzed."[26] Noting that the survey included only a small percentage of surgeons who used microsurgery (as does Kelman), and the superior results when microsurgery is used with existing procedures, Dr. Troutman believes that "should we repeat the study in 1977 when the majority of surgeons have adopted the microscope . . . we would probably find a clear advantage of intracapsular surgery over phacoemulsification."

Summarizing the advantages and disadvantages of his procedure at the Symposium on Eye Diseases of the Aged,[27] Dr. Kelman cited the advantages — the patient can be discharged the same or the following day and return immediately to all physical activities; patient can wear a contact lens in two to three weeks; it is the safest and easiest procedure for those under 40; wound complications are almost nil; and the elderly do not have to spend as much time in the hospital. Disadvantages — the technique is more difficult; surgeon must learn to use an operating microscope, which requires hours of practice; equipment is still costly (about $25,000); some surgeons

[26] September 27, 1976.
[27] Twenty-ninth Annual Meeting of the Gerontological Society, October 1976.

report that when a retinal detachment follows emulsification procedure, it is more difficult to repair (others disagree).

Dr. Kelman trains fifty physicians a month in a one-week course. Qualified graduates who have performed a hundred operations teach the procedure in other parts of the country. "The novice," Dr. Kelman advises, "should be assisted on his first patients, and trainees are urged to practice first on animals or eye-bank eyes before beginning on patients." Dr. Kelman is confident that as time goes by there will be more surgeons qualified in his procedure than not. Does that mean they will then adopt it?

A few weeks later, Johns Hopkins University's Dr. Charles E. Iliff, who uses both methods, shared his experiences with his colleagues at the 1976 AAOO national meeting. He finds the Kelman best for patients under 30. The machine, he says, is not only expensive to buy, it is also costly to maintain. More doctors in training, especially those taught to use the operating microscope in large institutions, should also be taught the use of the Kelman machine. They will then be better able to advise their patients which method of cataract extraction is best for them.

Dr. Iliff also points out that early discharge after surgery is neither new nor unique to Kelman. He himself keeps *all* his patients two days, and he cites reports of others, including a study of 1,000 consecutive cases done on an outpatient basis (1975) without an increase in complications.

As for early resumption of a normal life, that, Dr. Iliff contends, "depends on the patient's age and general physical condition. . . . The time of the return of the 40-year-old to bedroom sports, or the 90-year-old to being wheeled around his rose garden by an attractive attendant, does not depend on the method of cataract extraction, but on the desire and abilities of the patient to perform."

No matter what procedure is used, loss of the lens after cataract surgery means one-third less focusing power for all vision, but 100 percent loss for near vision. Reading glasses would take care of the near vision and a new prescription would take care of the rest, I reasoned. Such was my mood when I returned to the doctor to be examined for my first temporary glasses. I was elated when I read the chart with a new lens for the operated eye — I had forgotten how clear the letters could be. The doctor warned me I would need a period of adjustment. "For a while, it may be like walking with one shoe on and one shoe off." When I actually received my new glasses, I realized his warning was a gross understatement.

How much adjustment I was to need was evident to me when I stood at the window and looked down the street. "The new large street signs are a great improvement," I remarked. But then I noticed that the houses all appeared larger and closer. The children sat taller in the seats of their larger bicycles. It was an Alice-in-Wonderland effect. The distressing aspect of it all was that I now saw two images at the same time, and they failed to fuse.

I turned back into the room, removed the glasses in irritation, and hastily placed them on the table — or *thought* I did. They were actually an inch short of the table, and they fell to the floor. This was just one example of the clumsiness with which I was to perform the most elementary tasks.

Why did everything appear larger? Why did I see two different images? Why could I not judge simple distances? I learned that the eye from which the lens had been removed was now very farsighted. With that eye alone, everything appeared larger. The magnification of the corrective lens was, furthermore, much greater than that of the lens of the unoperated eye. In order for the brain to fuse the images from both eyes, the magnification should differ by no more than 10 percent. Because of the disparity between the two eyes, I had lost the fusion upon which I had depended all my life — and with loss of fusion I lost the ability to judge distance.

Now I understood why the doctor had predicted that I would ask for a second operation — an experience no longer fraught with the apprehension and uncertainty of the first. It was not until the second eye was done and I put on my first bifocal spectacles that I began to enjoy the full measure of clear, sharp and fused vision again.

Still, there are drawbacks. Because some side vision is lost, my family has learned not to approach me suddenly from the side, and when I collide with shoppers in a crowded store, I offer a hasty apology. For many, contact lenses are the answer — better cosmetically, better side vision, and better fusion, especially when only one eye has been operated on. The contact lens, because it is on the eye instead of one-half inch in front of it, provides less than 10 percent difference in magnification, and many who cannot tolerate the hard contact find help with the newer soft lenses. But there are still the feeble, the elderly, the arthritic, to whom handling the tiny lenses presents an insurmountable problem.

It has long been the dream of eye surgeons that someday they could promise their cataract patients a lens permanently placed in the eye to take over for the lens that has been removed. Credit for the first such attempt goes to an Italian surgeon in 1726, who used a glass lens. Maybe it was just as well that not much more was done for over two hundred years, until a plastic that promised to be safe and effective was developed. Much progress has been made in the last twenty-five years, and last year 10,000 patients in the United States received intra-ocular implants.

Hailed as a miracle by some and a menace by others,[28] it

[28] David Shoch, M.D., "The Intra-Ocular Lens — Miracle or Menace,"

is among the most controversial techniques today. The promise — vision closer to natural than either spectacles or contacts can provide; good peripheral vision; no fumbling with contacts which are easily lost, especially by the elderly; patients navigate well, even up and down steps. By 1977, more than 70,000 implants had been done in the United States. Why, then, does it remain controversial?

As the numbers of implants increased, so did reports on complications, sufficiently serious to prompt the House Subcommittee on Health, in 1975, to look into placing the lens under FDA regulation. At the time, no long-term testing of the safety of the material or the efficacy of the procedure had been done. It was available for sale to any surgeon who chose to use it. Testimony at the hearings included data from leading and highly respected ophthalmologists describing serious damage to eyes of many patients, glaucoma, corneal disease, inflammation and infection. In some instances, it was necessary to reoperate to remove the plastic lens because it had become displaced from its original location.

A large number of mishaps were (and continue to be) infectious, due to contaminated lenses and other defects traced to poor quality control by manufacturers, including shipments of lenses with sharp edges and others incorrectly labeled so that some patients received implants that made them unexpectedly nearsighted.

On November 18, 1977, the FDA[29] took a major step in safeguarding the welfare of tens of thousands of elderly for whom the intra-ocular implant can be closer to a miracle than a menace. Use of the intra-ocular lenses will be limited to investigational purposes. Only lenses that have been adequately tested and proved safe and effective will

---

The Sight-Saving Review, National Society for Prevention of Blindness, Summer 1976.

[29] Since May 1976, the lenses have been under the control of the FDA Medical Devices Division.

be approved for marketing. The first phase of the program will examine the records of about a dozen cooperating ophthalmologists who have performed lens implantation on large numbers of patients. It will look at the frequency of complications and at what factors were associated with success and failure, and will compare the outcome with alternative methods of restoring sight after cataract surgery.

Meanwhile, the AAOO, in a move to preserve the promise of the new advance and to avoid the pitfalls, established guidelines in 1975 for its use. Because present studies promise safety for no more than seven years, the recommendation is that it not be used in patients under 65 except in unusual circumstances.

Can you be *too* old for the implant? Brookdale Hospital Medical Center's Drs. Irwin Kanarek and Jacob Ackerman have just successfully removed a cataract and implanted an intra-ocular lens in an alert man of 101, almost totally blind in both eyes. "He is now back at home in familiar surroundings," Dr. Ackerman tells me. At 101, he is enjoying the benefits of restored vision.

Dr. Shoch's final word — "We are dealing not with a miracle, not with a menace, but with a reasonable advance in cataract surgery for certain selected patients." On the other hand, he is impressed with the large numbers of his patients who *do* adjust to spectacles. The key is to be "motivated and without infirmities," he says, which "probably describes 50 percent of our patients, men and women who put on their spectacles, grit their teeth, and proceed with life's tasks with little complaint."

My mother and I are among those who put on our spectacles and proceeded with life's tasks — not without complaints, I might add, but, on balance, with great satisfaction.

Among the numerous inquiries I received after my cataract story in *Family Circle* of June 1964 was one from a reader who wanted more information about my postoperative adjustment. She was writing on behalf of her dog, who was no longer catching a ball properly and displayed other signs of failing vision. The diagnosis is a cataract. The veterinarian is ready to operate. Do I think the dog would do well with contact lenses? Veterinarian Dr. William G. Magrane, author of *Canine Ophthalmology*, has a reassuring answer: After surgery, the dog can have pretty good distance vision even without contacts. At least two dog owners were so determined to restore optimum vision to their dogs that Dr. Magrane fitted them with contact lenses that they seem to be tolerating well. (Dogs most prone to cataracts are golden retrievers and beagles.) It is also reassuring that your dog has at least one other sense superior to ours that will help him — smell.

With cataracts, you have time on your side. The opacity may not get worse for years. Even then, you can make the decision for surgery only when it is serious enough to interfere with your day-to-day life. Not so with glaucoma, succinctly described by the National Society for Prevention of Blindness as "you don't feel a thing . . . after a while you don't see a thing." In the beginning, glaucoma gives no warning symptoms. Undetected, it can go on to cause irreversible damage, eventually total blindness — unnecessary because with early detection and prompt treatment, further damage can be halted and, in most instances, the threat of blindness eradicated. Indeed, glaucoma is an excellent candidate for a model of preventive medicine at its best.

There is no clear count of the number of glaucoma victims in the United States, but it has been estimated as high as 8 million, 2 to 3 million of whom are not aware that they have the disorder.[30] The Framingham Eye

[30] Dr. John Bellows, Director of the American Society of Contempo-

Study found a larger prevalence than expected — 3.3 percent — with a sharp increase in both men and women *over 65*, soaring from 1.4 percent in 52- to 64-year-olds to 7.2 percent in the 75- to 85-year range, with men in that group outnumbering women two to one.

While you are most likely to be at risk if you are over 38, you are at special risk if there is a history of glaucoma in your family. Over the years it has been observed that if one identical twin is affected, the other will be as well. Siblings tend to develop the disorder at about the same age, and it tends to run the same course for them. It may strike in successive generations — as many as three to five in some families — or it may skip a generation and reappear. The search for genetic markers has, until very recently, been unproductive. Now, new discoveries in genetics are yielding clues to heredity factors in glaucoma.

Washington University's renowned Dr. Bernard Becker turned his attention to the most complex genetic system in our makeup — HLA (human leukocyte antigens) — specific factors transmitted in our genes that make each of us unique and one of the liveliest areas of research today.[31] It is known that certain diseases are more common among individuals who possess certain factors than among those who do not. When they compared glaucoma patients to normal individuals, Dr. Becker and his colleagues found a significant increase in the prevalence of HLA B12 and HLA B7 among both black and white patients with glaucoma. They are now looking into the possibility that the presence of these HLA factors not only is a genetic marker, but may be of value in predicting who among patients with increased eye pressure will go on to develop

---

rary Ophthalmology and Professor of Ophthalmology, Chicago Medical School.

[31] See page 147, Chapter 4 for the HLA story — its role in juvenile diabetes, transplants, arthritis, and possibly cancer, multiple sclerosis and an increasing number of other disorders.

further damage.[32] An interesting link between HLA B12 and HLA B7 antigens and glaucoma comes from studies with Australian aborigines, among whom these antigens do *not* appear, and who have *no* glaucoma.

Glaucoma starts with an increased pressure in the clear circulating fluid in the front of your eye — the aqueous humor — whose function is to provide nourishment for the cornea and the lens. Normally, new fluid is constantly being formed and drained off through a tiny drainage system. When this outflow is impaired, pressure begins to build up. Unchecked, it can cause a loss of visual acuity (some victims change their glasses two or three times in a relatively short period, but to no avail); changes in field of vision (restricting vision to only what is directly in front of you); and, at worst, damage to the optic nerve. In the most common form—chronic primary open-angle glaucoma—it can all happen without any pain or other warning signs.[33]

A rarer form, acute glaucoma, often starts with pain severe enough to cause nausea and vomiting. In persons born with a specific anatomical structure — a narrow angle between the cornea and iris — the glaucoma can be triggered by emotional stress.[34] If ignored, blindness may occur with extraordinary rapidity. With early diagnosis, a permanent cure can be achieved with surgery.

Your eye doctor will make a diagnosis of glaucoma (chronic open-angle) only after very careful scrutiny. In addition to taking repeated eye-pressure measurements (it fluctuates from hour to hour, month to month, season to season), he will test your side vision (visual field) and take

[32] Archives of *Ophthalmology*, February 1977.

[33] Halos around lights, excessively long time to readjust from light to dark, blurred vision, headaches, etc.

[34] Dr. Jules François, international authority, says most susceptible is the "nervous, emotional, anxious female before or at menopause." But men are not immune, also often associated with deep emotional upheavals.

a good look at the back of the eye, specifically the disc — the spot where the optic nerve exits from the retina and goes to the brain. Here is where changes that may signal future blindness take place.[35]

Loss of vision suffered before the diagnosis is made cannot be restored, but further deterioration can be halted. The mainstay of treatment is pilocarpine, a substance that reduces pressure by promoting the outflow of the fluid.[36] While it is safe and effective, it has disadvantages — side reactions, and the fact that drops in the eyes must be administered every six hours.

Now a new drug related to a family of drugs causing much excitement in heart disease and high blood pressure — beta blockers — is causing excitement in glaucoma as well. Available since October 1978, it seems to be free of side effects and to be effective with only two drops a day — one in the morning and one at night — a special convenience to the older patient. "Timelol may cause us to take a new look at glaucoma therapy," says Dr. Thom J. Zimmerman, the young ophthalmologist who conducted the early studies.[37]

To Drs. Robert S. Hepler and Ira Frank of U.C.L.A. goes the credit for treating glaucoma with the most con-

---

[35] The central portion of the disc is a cup-shaped funnel-like space. With increased eye pressure from untreated glaucoma, the entire disc can become depressed and cup-shaped. It is important for the doctor to follow the cup-disc ratio in treatment of glaucoma. An increasing number of ophthalmologists are taking periodic stereo photographs, enabling them to follow changes without depending on memory. If the treatment is working, there should be no changes. A major goal for prevention is to develop a technique to measure cup-disc changes *before* loss of vision develops. Dr. Liebowitz and physicist Gerald Shapiro at Boston University are now working on a procedure to obtain a computer map of the optic disc, promising to warn objectively of trouble ahead.

[36] Sometimes it is necessary to prescribe a medication that reduces production of the fluid.

[37] *Medical World News*, March 7, 1977.

troversial drug of the twentieth century — marijuana. Used therapeutically for almost 4,000 years until it was abruptly removed from the United States Pharmacopeia in 1941, it is now beginning to reemerge as a "healer." Dr. Keith Green of the Medical College of Georgia recalls the "skepticism, mirth and bad publicity"[38] that surrounded Dr. Hepler's first report in *JAMA* in 1971. His subjects were young men smoking a high grade of pot supplied by the NIMH. Their intra-ocular pressure was reduced by as much as 45 percent.

Undeterred by professional scorn and public outrage, the U.C.L.A. group continued their investigation, and were joined by Dr. Green and others, who confirmed the effects in rabbits and humans. Both marijuana and Δ^9 THC (an active ingredient) bring pressure down in normals and in glaucoma victims, but the two routes used — smoking and injections — are not suitable for sustained therapy. Work is in progress now to develop a preparation that can be applied directly to the eye, is effective and safe for long-term use, and is free of undesirable side effects. Oddly enough, a few of the older patients in the U.C.L.A. study described the high so popular with recreational users as an "undesirable side effect."

By 1978, the emotional and political climate in regard to marijuana had changed considerably. Medically, cannabis may be on its way back to the medicine cabinet — for glaucoma; to relieve nausea of cancer chemotherapy; as an anti-anxiety sedative; to reduce migraine; and to protect against asthma. How much has changed in less than half a dozen years was highlighted when the New Mexico Legislature passed a bill in February 1978 authorizing research with marijuana and to contract with the National Institute on Drug Abuse to make doses available to qualified patients.

The growing concern today with glaucoma was re-

[38] *Investigative Ophthalmology*, April 1975.

flected by the attendance of 1,500 United States ophthalmologists at the first International Glaucoma Congress in
1977, and by popular request of the participants, the
Second International Glaucoma Congress was held
exactly one year later in 1978.

The successes scored in saving sight in the aging have
not, unfortunately, been matched by progress in their
total health care. There are few disorders unique to the
elderly, but what is different is often the way they show
up[39] — a coronary attack without chest pain, appendicitis
without abdominal tenderness, infection without fever,
and even an overactive thyroid unexpectedly accompanied
by loss of appetite and apathy. That practicing physicians
themselves are beginning to acknowledge their limitations
in dealing with such situations was underscored in a recent
survey conducted by an AMA news periodical.[40] When
asked, "Do MD's need special training in geriatrics?" an
impressive 75 percent answered "Yes."

There are also indications that medical students are
developing a greater sensitivity to the needs of the aged. In
contrast to attitudes revealed in a study at a West Coast
medical school ten years ago, when "crocks," "biddies,
over the hill" were among the descriptions used, the
American Medical Student Association today is asking that
geriatric medicine be included in the curriculum.

It won't be easy for geriatric medicine to come of age in
the United States. "Current estimates are that perhaps less
than 15 of an estimated 25,000 faculty members of American medical schools have any genuine expertise."[41] Dr.

[39] *Cowdry's The Care of the Geriatric Patient*, fifth edition, edited by F. U.
Steinberg (St. Louis: C. V. Mosby, 1976).

[40] *Impact*, September 1976.

[41] The first chair of geriatric medicine in the United States was established by the Cornell Medical Center–New York Hospital in 1977.
By contrast, England has ten, Sweden two, Holland one. There are

Butler testified before the United States Senate Special Committee on Aging: "The question is, how do you create eggs without chickens?"

Perhaps the most serious indignity inflicted on the elderly is the glib diagnosis of "senility," with its lifetime sentence of irreversible and hopeless organic brain damage. But not every person who can't remember what he had for breakfast (or whether he had breakfast at all), can't add up the items on his brief shopping list, and then gets lost on the short walk home is senile. He may be suffering from malnutrition, anemia, too many or poorly prescribed doses of drugs,[42] or depression.[43] *All can be diagnosed; all can be treated; many can be reversed.*

Nevertheless, there are still hundreds of thousands with organic brain disease in whom the progressive deterioration continues inexorably. They comprise half the nursing home population and account for 120,000 deaths a year.

---

only two residencies in the United States, both established by Dr. Leslie S. Libow — the first at Mt. Sinai City Hospital Medical Center, Elmhurst, New York, in 1972, and the second when he moved to the Jewish Institute for Geriatric Care, Long Island Jewish–Hillside Medical Center, New Hyde Park, New York. The Veterans Administration has also started a program to train doctors in geriatrics, anticipating the increased needs for geriatric services for the 13 million World War II veterans in or nearing their sixties.

In a communication to the *New England Journal of Medicine*, April 6, 1978, Dr. Libow reports that ". . . geriatric medical programs for students and for resident physicians exist or are being established at the following medical schools: New York University, University of Washington in Seattle, University of California at Los Angeles, and the universities of Arkansas, North Dakota, Duke, Harvard, Stanford, Wisconsin, Rochester, Illinois, Maryland and Boston."

[42] Very little is known about why the elderly often react to drugs with unexpected and adverse effects. To help establish prescription guidelines, the NIA has awarded a contract to the Boston Collaborative Drug Surveillance Program.

[43] The over-65 account for 25 percent of all suicides, but they represent only 2 percent of those seen in community mental health centers.

While scientists are agreed that senile dementia *is not an inevitable consequence of aging*, they have had, until very recently, virtually no clues to why some persons are affected and others escape. There are, however, some degenerative disorders with early symptoms much like senility that strike people in their forties and fifties, and among these pre-senile dementias, Alzheimer's disease is the most common. Autopsy studies of the brains of most senile patients show characteristic degeneration identical with that found in Alzheimer's disease![44]

Dramatic discoveries in the last few years may, for the first time, shed light on the problem of senility, which affects no fewer than 600,000 persons and possibly as many as a million. The key words are "slow-growing virus," and the event that alerted scientists concerned with senility (senile dementia) is the research that won the 1976 Nobel prize for Dr. Carleton Gajdusek. The NIH scientist had long been interested in Kuru, a mysterious fatal brain disorder that claimed the lives of many members of a cannabalistic tribe in New Guinea. Because it often affected families, it was believed to be genetic. Dr. Gajdusek, in the best detective tradition, had another thought. It was the custom of the tribe for surviving members of a family to consume some of the deceased's brain. They did not get the wisdom they sought. On the contrary, they became infected with the same slow-growing virus that had killed their late relative.

Later, Dr. Gajdusek and his colleagues studied another rare degenerative brain disorder, Creutzfeldt-Jacob disease. Suspecting that an infectious agent — a very slow-growing virus — was the culprit, they injected material from the brains of human victims into laboratory animals. After many months, the animals developed the same

[44] Dr. Robert Terry, Albert Einstein College of Medicine, New York City.

symptoms. The diseases had been transmitted — Kuru and Creutzfeldt-Jacob disease are infectious in nature. Slow viruses are now suspect in a number of other disorders of the nervous system, such as Parkinson's disease, multiple sclerosis and the most recent under scrutiny, Alzheimer's disease, with its striking similarity to senile dementia.

In June 1977, the NIH brought together distinguished scientists from around the world to review the current knowledge on the dementias, and to help chart a course for future research.[45] Among the highlights of the conference was evidence linking senile dementia to Alzheimer's and Alzheimer's to a slow virus; observations on how degeneration of specific nerve cells disrupts transmission of chemical messages in the brain. Still another study just in an early preliminary stage suggests that by enhancing a specific neural transmission system, memory and thinking capacity can be improved in the aged.

Scientists of the three NIH agencies sponsoring the conference[46] place a high priority on solving the problems of senility, not only because of today's needs, but for the future as well. Today, the majority of the old are the "young-old" — over 65 and not yet 75 — but the number of "old-old" is rising. In fact, the over-85 are the fastest growing segment of the population. By the year 2000, there will be 154 women to every 100 men,[47] and with senility more common in women than men, it will be a dismal scene indeed. It will be a grim irony if middle age is successfully stretched to old age only to have the last years spent stripped of intellect and totally dependent.

---

[45] *NIH Record*, July 12, 1977.

[46] National Institute of Neurological and Communicative Disorders and Stroke (NINCDS); National Institute on Aging (NIA); and the National Institute of Mental Health (NIMH).

[47] Men die at an earlier age not only for some still unexplained biologi-

On the other hand, there is a strength and vitality in many of today's "young-old" that may help stave off senility. Nor is "young-old" rigidly chronological. Walter, Uncle Louie and Ann attest to that. The "young-old" are better educated, have lived and fought through the depression and the war, have helped organize unions and political movements, have even survived the experience of "flower children" offspring.

"We are recycling ourselves," says 70-year-old Doris Mendiz at the 1977 convention of the Gray Panthers, a small but determined action group working on guidelines for health care services (Medicare has some glaring omissions — out-of-pocket costs are two to three times as high as medical expenses were before Medicare; eyeglasses and dental services are not covered; out-of-hospital drugs can be beyond the reach of many, especially the large number of individuals with lifetime needs for treatment of high blood pressure, diabetes, heart disease, glaucoma, etc.).

The Gray Panthers went to an AMA convention to let the doctors know that it might be a good idea to start making house calls again; they succeeded in getting the loan policy in Philadelphia banks improved; they are vocal in seeking a better image for the elderly on TV and in other media; and they fought vigorously for a higher retirement age. Says Panther founder, 73-year-old Maggie Kuhn, "We've experienced the Black Movement and the Women's Movement; now it will be the Wrinkled Radical Movement — and watch out, because we come with years of experience behind us."[48]

Some "young-old" are forging new careers. In her first bid for public office, 73-year-old widow Isabella Cannon

cal reasons, but also from behavior that courts early death — more violence, guns, speeding, drinking, etc.

[48] *New York Times*, October 3, 1977.

defeated the incumbent in the mayoralty race in Raleigh, North Carolina. The major issue, shared with many American cities in 1977, was the city's economic future. "I owe my success to a coalition that cut across age, race and economic lines," Mayor Cannon told the *AARP News Bulletin*.[49] "Old people have two priceless assets — time and a wealth of experience. Public service is a good field for sharing those gifts. I strongly recommend it to men and women of my age."

The world is being let in on a secret it preferred not to know — that sex does not stop at 60. Two of today's leading gerontologists have written about the subject recently with warmth, perception and knowledge. Dr. Butler[50] chose the subject as a change of pace after the somber 1975 *Why Survive? Being Old in America*. Alex Comfort, described by a fellow scientist as having turned from Methuselah to Eros, makes a beautiful presentation not just of sex, but of what aging *can* and *should* be in general, in *A Good Age*.[51]

The aging and aged today are not only more visible and more vocal; they are also more varied, both in biology and behavior. Finding a suitable label for all is becoming increasingly difficult. "Over 65" is a useful administrative category; "Golden Age" was tried and discarded — it was too obviously a meaningless euphemism; "Senior Citizen" — who are the junior citizens?

Some refuse to be locked into any label whatsoever. Said sculptor Jacques Lipchitz, in a documentary on his life and work, "I am not 80, I am four times 20."

---

[49] American Association of Retired Persons.
[50] Robert N. Butler, M.D., and Myrna I. Lewis, ACSW, *Sex After Sixty—A Guide for Men and Women in Their Later Years* (New York: Harper & Row, 1976).
[51] New York: Crown Publishers, 1976.

Suggestions for Further Reading

Understanding Heredity

Apgar, Virginia, and Beck, Joan. *Is My Baby All Right? A Guide to Birth Defects*. New York: Trident Press, 1972. For birth defects — both hereditary ones and those associated with pregnancy and delivery.

Dobzhansky, Theodosius. *Mankind Evolving*. New York: Bantam Books, 1970.

Etzione, Amitai. *Genetic Fix*. New York: The Macmillan Company, Inc., 1974.

Greenblatt, Augusta. *Heredity and You: How You Can Protect Your Family's Future*. New York: Coward, McCann & Geoghegan, Inc., 1974.

Grobstein, Clifford. "The Recombinant-DNA Debate," *Scientific American*, July 1977.

Lurie, S. E. *Life, the Unfinished Experiment*. New York: Charles Scribner's Sons, 1973. A Nobel Laureate interprets modern biology.

Peters, James A. *Classic Papers in Genetics*. Englewood Cliffs, N.J.: Prentice-Hall, Inc., 1959.

Sayre, Anne. *Rosalind Franklin and DNA*. New York: W. W. Norton & Co., Inc., 1975. A vivid view of what it is like to be a gifted woman in an especially male profession.

Coronary Heart Disease, as well as high blood pressure, obesity and aging.

Eating Right for Less. Consumers Union, Orangeburg, N.Y. 10962. Updated 1977. Guide to wholesome, low-cost nutrition for older people.

Eshleman, Ruth, and Winston, Mary. *The American Heart Association Cookbook.* New York: David McKay Company, 1975.

Mayer, Jean. *A Diet for Living.* New York: David McKay Company, 1975. Paperbound, Consumers Union, Orangeburg, N.Y. 10962.

Diabetes

Diabetes Forecast. A bimonthly publication of the American Diabetes Association, 1 West 48th Street, New York, N.Y. 10020.

Aging

Butler, Robert N., M.D. *Why Survive? Being Old in America.* New York: Harper & Row, 1975.

Butler, Robert N., M.D., and Lewis, Myrna I., ACSW. *Sex After Sixty – A Guide for Men and Women in Their Later Years.* New York: Harper & Row, 1976.

Comfort, Alex, M.D. *A Good Age.* New York: Crown Publishers, 1976.

Otten, Jane, and Shelley, Florence D. *When Your Parents Grow Old.* New York: The New American Library, Inc., 1978.

Rosenfeld, Albert. *Prolongevity.* New York: Alfred A. Knopf, 1976.

Seidman, Lloyd. *New York City: Retirement Village.* New York: Harper & Row, 1977. A handy reference of resources for the retired and those who will be retired.

Where to find Help

Your local Heart Association has a variety of publications on recipes, diet, etc., for you, your children, your children's teachers, your doctor. For a local address, write to American Heart Association, 7320 Greenville Avenue, Dallas, Texas 75231.

When you have a special problem with cholesterol and other fats:

For Your Doctor:
Dietary Management of Hyperlipoproteinemia. A Handbook for Physicians and Dietitians. Prepared and compiled under the direction of Donald S. Fredrickson, M.D., Robert I. Levy, M.D., Miss Merme Bonnell, R.D., Mrs. Nancy Ernst, R.D. Also available to your doctor are appropriate diets. Office of Information, National Heart, Lung and Blood Institute, National Institutes of Health, Bethesda, Md. 20014

National High Blood Pressure Information Center 120/80
National Institutes of Health
Bethesda, Md. 20014

American Diabetes Association, Inc.
One West 48th Street
New York, N.Y. 10020

Juvenile Diabetes Foundation
23 East 26th Street
New York, N.Y. 10010

National Foundation — March of Dimes
Box 2000
White Plains, N.Y. 10602

National Genetics Foundation
250 West 57th Street
New York, N.Y. 10019

National Society for the Prevention of Blindness, Inc.
79 Madison Avenue
New York, N.Y. 10016
(Saving your sight is covered in Chapter 4 on Diabetes and
    Chapter 6 on Aging.)

Index

Ackerman, Jacob, 196

Aging, xiii, xiv, 172–208; cataracts and, 185, 186, 188–97; cell research and, 181–84; diabetic retinopathy and, 185–87; Empathetic Model and, 184–85; endocrine system and, 180–81; genetic programming and, 176; glaucoma and, 185, 186, 197–202; immune system and, 178–79; macular degeneration and, 185, 187, 188; "senility" and, 203–6; sex and, 207; stereotypes and, 174–75; suicide and, 203n; thyroid activity and, 181; total health care and, 202–3

Ahlquist, Raymond P., 108n

Ahrens, Edward, 44, 75

Alcohol: CHD and, 76; hypertension and, 92

Alderman, Michael H., 111

Allen, Woody, 1

Alphafetoprotein (AFP), 20, 21

Alrestatin, 134

Alzheimer's disease, 204, 205

American Academy of Ophthalmology and Otolaryngology (AAOO), 191, 196

American Academy of Pediatrics, 29

American Association for the Ad-

vancement of Science (AAAS), 103

American Association of Retired Persons (AARP), 207

American College of Cardiology, 59

American Diabetes Association, xvi, 115, 134, 135, 140, 152

American Health Foundation, 29, 38–39, 76

American Heart Association, 29, 30, 35–36, 54, 86–87, 106; Science Writer's Forum of, 29–30, 38, 48–49, 136

American Medical Association (AMA), 202, 206

American Medical Joggers Association, 62

American Medical Student Association, 202

American Society of Contemporary Ophthalmology, 197n

Amino acids, 10–11

Amniocentesis, 18–21; Down syndrome, 19–20; familial hypercholesterolemia, 21; neural tube defects, 20–21; Tay Sachs, 18. *See also* Genetic counseling

Angina, 60

Angiography, 78; fluorescein, 187n

Angiotensin II, 108

Anitschkow, N. N., 45

Anti-platelet serum, 69–71
Arber, Werner, 8n
Arcus cornea, 32
Armstrong, M. L., 68n
Arnold, Matthew, 4
Arnold, Thomas, 4–5
Arrhythmia, 53, 55
Arteriosclerosis, 26
Arthritis, obesity and, 157
Aruacana fowl, 61
Aryl hydrocarbons, 71–72
Aspirin: blood platelets and, 77,
    128; CHD and, 77; strokes and,
    77; in treatment of retinopathy,
    127–28
Atheromatous plague, 68–71
Atherosclerosis, xiii; in antiquity,
    25–26; definition of, 24n; high
    blood pressure and, 86–87; pre-
    vention of, 29, 35–42, 45, 72–78;
    research on origins of, 45–52,
    68–72; reversal of, 66–68; Task
    Force on Genetic Factors in Ath-
    erosclerotic Disease, 29. See also
    Coronary heart disease
Australian aborigines, 199
Autonomic nervous system, volun-
    tary control of, 101–6
Avery, Oswald T., 8

B cells, 178
Baby food, 96–97
Baer, John E., 106n
Baltimore, David, 8n
Banting, Frederick, 119
Banting, William, 166–67
Baro-receptors, 85
Barr, David, 48
Bassler, Thomas J., 63, 64
Beadle, George, 8n
Becker, Bernard, 198
Behavior modification, obesity and,
    170–71
Bellows, John, 197n
Benditt, Earl P., 71–72
Benditt, J. M., 71
Benson, Herbert, 59–60, 64, 99–
    100, 103–5
Berg, Paul, 2n
Berson, Samuel, 120n

Best, Charles, 119
Beta blockers (beta adrenergic block-
    ing agents): in treatment of an-
    gina, 60; glaucoma, 200–1; high
    blood pressure, 107–8
Beyer, Karl H., Jr., 106n
Bierman, Edwin L., 31
Biofeedback, hypertension and,
    100–6
Biron, Paul, 81
Black, James W., 108n
Blackburn, George L., 168n
Blacks: diabetic mothers among,
    150; high blood pressure among,
    xiii, 4, 82–83, 91, 109–11. See also
    Sickle cell disease
Blankenhorn, David, 66, 67
Blecher, Melvin, 125
Blood pressure, high, see High blood
    pressure
Blood sugar, 118–25, 134–40; com-
    plications of diabetes and, 134–
    36; glucagon and, 136–37; mea-
    surement of, advances in, 135–36;
    somatostatin and, 138–40
Blumenthal, Sidney, 80, 90
Boston Collaborative Drug Surveil-
    lance Program, 77, 203n
Bowie, E. J. W., 70
Brady, Roscoe D., 18
Brain hormones, 101n
Bray, George A., 155, 166
Brazeau, Paul, 139
British Medical Journal, 58, 79,
    105–6, 126, 129
Brookdale Hospital Medical Center,
    196
Brown, Michael S., 21, 33, 47
Bruenn, Howard G., 112
Buchwald, Henry, 67
Burton, Dee, 76
Butler, Robert N., 174–75, 177, 207

Cahill, George F., Jr., 115, 135,
    139, 148
California, State Education De-
    partment of, xiv
Cancer, T cells and, 179–80
Canine Ophthalmology (Magrane), 197
Cannabis: diabetes and, 152–53;

Wilson's disease, xiv
Winikoff, Beverly, 74*n*
Wissler, Robert W., 28*n*
Wodinsky, Harvey, 176
Women: acute glaucoma in, 199*n;*
 CHD in, 34*n*, 50, 57–58, 76, 78,
 132, 157; hypertension in, 91–92;
 senility in, 206; smoking among,
 37, 92
Wood, Peter, 50
Woodrow, J. C., 147
World Health Organization, 26, 29,
 73

Wynder, Ernst L., 29, 72*n*

Xanthoma, 32, 33

Yalow, Rosalyn, 120
Yoga, 100, 102, 103

Zen communes, study of, 50
Zimmerman, Thomas J., 200–1
Zinc, 79
Zinner, Stephen H., 80*n*

About the Authors

AUGUSTA GREENBLATT is a clinical scientist and a nationally renowned medical and science writer. She is the author of *Heredity and You: How You Can Protect Your Family's Future; Teen-age Medicine: Questions Young People Ask About Their Health;* and *Why Do I Feel This Way?* (paperback revised version of *Teen-age Medicine*). Her work has been acclaimed by physicians, researchers and educators in leading universities for its timeliness, accuracy and clarity. She has also received high praise from reviewers in newspapers and magazines across the country for her engrossing narrative style.

Since 1964, when Mrs. Greenblatt switched careers from the laboratory to the typewriter and lectern, she has contributed to popular national magazines, including *Family Circle, Woman's Day, McCall's, Parents, Coronet, Grolier's Encyclopedia Science Supplement*. Reprints of her magazine articles have been used by physicians for their patients, and her books are in use in high schools and colleges. Her professional publications include contributions to the *Journal of Biochemistry, Archives of Biochemistry, American Journal of Medical Sciences, Army Medical Bulletin, Transactions of the American Chemical Society,* and *Adult Leadership*.

Mrs. Greenblatt is a graduate of Cornell University, studied biochemistry in the graduate division of Brooklyn College, CUNY, and received an M.S. from Hofstra University. She is certified to teach all sciences in secondary schools. She was formerly a writer-editor for the United States Public Health Service; supervisor of clinical chemistry and microbiology in the New York City Department of Hospitals; civilian supervisor of clinical laboratories in the Army Medical Department; and Director of a Clinical Diagnostic Laboratory. More recently, she has been a lecturer at the New York University School of Continuing Education, and her daytime course, "What's Behind the Headlines in Science," has attracted several hundred women (plus a few men) in the Hewlett-Woodmere Continuing Education Program. She is a member of the American Public Health Association, National Association of Science Writers, and American Medical Writers Association.

Your Genes and Your Destiny is the first joint writing venture for Dr. and Mrs. Greenblatt since their research publications in the early days of their marriage.

I. J. GREENBLATT, Ph.D., has been in the field of laboratory medicine for over forty years and has recently retired from the position of Director of Clinical Laboratories, Brookdale Hospital Medical Center, Brooklyn, New York. Dr. Greenblatt was one of the early pioneers in the study of atherosclerosis and coronary heart disease, has published more than fifty papers in medical and scientific journals, including a chapter in *A Pharmacological Approach to the Study of the Mind* (edited by Featherstone and Simon), Charles C Thomas, 1959. His major research interests, in addition to atherosclerosis, are clinical pharmacology and toxicology.

He has headed up research teams that won two awards from the American Medical Association: one for a study of the use of ion exchange polymers to remove excess sodium, and the other for one of the early studies on the relationship between blood fats, atherosclerosis and coronary heart disease.

Dr. Greenblatt received his B.A. at New York University, M.S. and Ph.D. at Georgetown University. During World War II he was a Captain in the United States Army, serving first as Chief of the Laboratory Service, Station Hospital, Camp

226 ABOUT THE AUTHORS

Stoneman, California, and then overseas in the Pacific theatre. He was one of the first American scientists to enter Hiroshima.

He is certified by the American Board of Clinical Chemistry, a Fellow of the National Academy of Clinical Biochemistry, the American Institute of Chemists, and the Association for the Study of Arteriosclerosis (American Heart Association); a member of the Society of Experimental Biology and Medicine, the Association of Clinical Scientists, the American Chemical Society, the American Association for Advancement of Science; and a life member of the New York Academy of Sciences.

His academic experience includes: Lecturer — Clinical Biochemistry, Long Island College of Medicine (now SUNY Downstate); Adj. Associate Professor, Occupational Medicine, Columbia College of Physicians and Surgeons — School of Public Health; Clinical Associate Professor of Pathology, New York University College of Medicine.

At the present time, Dr. Greenblatt is Associate Professor of Clinical Pathology at the Arnold and Marie Schwartz College of Pharmacy and Health Sciences of Long Island University.

Dr. and Mrs. Greenblatt live in Woodmere, New York.